Hybrid Nanomaterials - Flexible Electronics Materials

Edited by Rafael Vargas-Bernal, Peng He and Shuye Zhang

Published in London, United Kingdom

IntechOpen

Supporting open minds since 2005

Hybrid Nanomaterials - Flexible Electronics Materials
http://dx.doi.org/10.5772/intechopen.83326
Edited by Rafael Vargas-Bernal, Peng He and Shuye Zhang

Contributors
Gen Long, Ning Wang, Xiaolong Chen, Shuigang Xu, GuiPing Dai, Hua-Fei Li, Shuguang Deng, Saleem Khan, Shawkat Ali, Amine Bermak, Arun Dev Dhar Dwivedi, Sushil Kumar Jain, Rajeev Dhar Dwivedi, Shubham Dadhich, Hyun-Joong Kim, Jung-Hun Lee, Ji-Soo Kim, Tae-Hyung Lee, Lidong Chen, Xun Shi, Tian-Ran Wei, Heyang Chen, Amita Somya, Rafael Vargas-Bernal

First published in London, United Kingdom, 2020 by IntechOpen
IntechOpen is the global imprint of INTECHOPEN LIMITED, registered in England and Wales, registration number: 11086078, 7th floor, 10 Lower Thames Street, London, EC3R 6AF, United Kingdom
Printed in Croatia

British Library Cataloguing-in-Publication Data
A catalogue record for this book is available from the British Library

Additional hard and PDF copies can be obtained from orders@intechopen.com

Hybrid Nanomaterials - Flexible Electronics Materials
Edited by Rafael Vargas-Bernal, Peng He and Shuye Zhang
p. cm.
Print ISBN 978-1-83880-337-7
Online ISBN 978-1-83880-338-4
eBook (PDF) ISBN 978-1-78985-651-4

For EU product safety concerns:
IN TECH d.o.o., Prolaz Marije Krucifikse Kozulić 3, 51000 Rijeka, Croatia, info@intechopen.com or visit our website at intechopen.com.

Meet the editors

Rafael Vargas-Bernal received his DSc in Electronic Engineering from the National Institute of Astrophysics, Optics and Electronics (INAOE), Tonantzintla, Puebla, Mexico, in 2000. He was associate professor in the Department of Electronics Engineering from 2002 to 2010, and since 2010 he has held the same position in the Materials Engineering Department at the Instituto Tecnológico Superior de Irapuato (ITESI), Irapuato, Guanajuato, Mexico. He is a level-1 researcher of the National System of Researchers from Mexico. He also belongs to the Advanced Materials Applied to Engineering research group called and has PRODEP (Program for the Professional Development of Teachers) desirable profile. He has been a reviewer for journals for RSC, Elsevier, and IEEE. He has published thirtten articles in indexed journals and thirty book chapters. His areas of interest are nanomaterials, two-dimensional materials, graphene, composites, biosensors, and gas sensors.

Prof. Peng He is the Duty Director of the State Key Laboratory of Advanced Welding and Joining, China. He has been an Executive Member of the ISO-Welding Standard Committee from 2014 and has been an Executive Member of the International Materials Science & Joining Technologies Committee from 2012. Up to now, he has written 292 papers and 7 books, developed 39 patents, and received 13 awards in materials science. He is currently researching materials processing, micro-joining, and reliability.

Dr. Zhang has been a lecturer at State Key Lab of Advanced Welding and Joining since 2017. He is focusing on novel and advanced materials for nanofabrication and nanodevices. He has published forty SCI&EI indexed papers, including ten IEEE conference papers, in IEEE Transactions on Components, Packaging and Manufacturing Technology, Journal of Materials Science: Materials in Electronics, Journal of Alloys and Compounds, and Advanced Materials. Dr. Zhang is the session chair of IEEE NANO 2019 and co-session chair of IEEE 3M NANO 2018 and 2019.

Contents

Preface

Two of the hottest research topics today are hybrid nanomaterials and flexible electronics. Both depend on advances in materials engineering that researchers make at theoretical, modeled, and experimental levels. This book covers some of the advances made in the last decade regarding engineering materials and their applications, particularly in the area of electronics.

This book presents different research works dedicated to the study of hybrid nanomaterials and materials for flexible electronics developed by researchers around the world.

The book is divided into two parts, with Chapters 1 to 4 belonging to the first part on "Hybrid Nanomaterials," and Chapters 5 to 8 belonging to the second part on 'Flexible Electronics."

In Chapter 1, "Introductory Chapter: Hybrid Nanomaterials" by Rafael Vargas-Bernal, the author presents an updated introduction to hybrid nanomaterials and their recent advances from this decade. Although many publications have been developed by researchers around the world, the extraordinary physical and chemical properties of these types of materials, as well as the products derived from them, will continue to amaze us in this century.

In Chapter 2 "Electronic Transport in Few-Layer Black Phosphorus" by Gen Long, Xiaolong Chen, Shuigang Xu, and Ning Wang, the authors analyze the electron transport properties of phosphorene with few layers through the Landau level model, two-dimensional limit, magnetoresistance, and Hall quantum effects such as Shubnikov-de Hass (SdH) oscillations. This study exemplifies its application for the design of a phosphorene-based field effect transistor (FET).

In Chapter 3 "Synthesis of Three-Dimensional Nanocarbon Hybrids by Chemical Vapor Deposition" by Hua-Fei Li, Shuguang Deng, and Gui-Ping Dai, the authors highlight the synthesis, unique properties, and applications of three-dimensional nanocarbon hybrids synthesized using deposition of chemical vapor in the field of electronics. These materials have advantages over other types of carbon-based nanomaterials.

In Chapter 4 "Hybrid Ion Exchangers" by Amita Somya, the author discusses the advantages of metal phosphate-based hybrid ion exchangers using surfactants or surface-active agents. These allow an increase in ion exchange capacity for industrial and environmental applications such as water pollution control. Furthermore, the author also examines the physical and chemical properties of these materials.

In Chapter 5 "Pressure-Sensitive Adhesives for Flexible Display Applications" by Tae-Hyung Lee, Ji-Soo Kim, Jung-Hun Lee, and Hyun-Joong Kim, the authors discuss the development and performance of electronic systems that require the use of pressure-sensitive adhesives (PSA) and optically clear adhesives (OCA). These adhesives, in addition to providing mechanical fastening, must have multiple physical functionalities to be used on flexible displays in order to protect vulnerable electronic devices.

In Chapter 6 “Numerical Simulation and Compact Modeling of Thin Film Transistors for Future Flexible Electronics” by Arun Dev Dhar Dwivedi, the author introduces the importance of numerical simulation and compact modeling in SPICE, a general-purpose, open-source analog electronic circuit simulator, in the design of organic thin film transistors (OTFTs) for flexible electronics.

In Chapter 7 “Smart Manufacturing Technologies for Printed Electronics” by Saleem Khan, Shawkat Ali, and Amine Bermak, the authors review the different techniques used in printed electronics based on technology of assembly by roll-to-roll systems to implement electronic devices and circuits. Furthermore, they highlight possible alternatives and potential research directions related to this area.

In Chapter 8 “Plastic Inorganic Semiconductors for Flexible Electronics” by Tian-Ran Wei, Heyang Chen, Xun Shi, and Lidong Chen, the authors highlight the relevance of applying Ag2S as a ductile inorganic semiconductor by correlating its chemical and physical properties for applications in flexible electronics and in particular to develop thermoelectric devices by alloying sulfide with either selenium and/or tellurium.

Considering the currently high level of interest in hybrid nanomaterials and flexible electronics, this book is a valuable source of information for materials scientists and engineers working in the field as well as for graduate students in materials engineering working in the area of electronics.

We wish to thank the contributing authors as well as the staff at IntechOpen, especially Ms. Rebekah Pribetic, for their invaluable assistance during the book production process. Finally, I wish to thank my wife and son for their help and support through this endeavor. My best wishes to all the authors and readers of this book.

Dr. Rafael Vargas-Bernal (Corresponding editor)
Full Professor,
Department of Materials Engineering,
Instituto Tecnológico Superior de Irapuato (ITESI),
Irapuato, Guanajuato, México

Dr. Peng He and Dr. Shuye Zhang (Co-editors)
Harbin Institute of Technology,
China

Section 1

Hybrid Nanomaterials

Chapter 1

Introductory Chapter: Hybrid Nanomaterials

Rafael Vargas-Bernal

1. Introduction

One of the most exciting research fields in recent decades in the area of materials engineering is that of hybrid nanomaterials. These materials possess extraordinary physical and chemical properties derived from their size in the nanoscale. Among the reasons for this technological and scientific trend are its multidisciplinarity and the combination of the best attributes of both inorganic and organic chemistry, which give rise to multifunctional materials based on an approximation of building blocks. In addition, there is the possibility of incorporating the physical and biological sciences to produce biomimetic approaches to create unique materials derived from the requirements of emerging technologies that lead to the development of a current driving force to perform unprecedented research of materials, devices, and applications. So far, although many reviews, articles, and books are being continuously published, the scientific literature continues to surprise with new contributions and different views of researchers around the world. The purpose of this chapter is to present an updated introduction of hybrid nanomaterials and their recent advances of this decade.

2. Basic concepts

Hybrid nanomaterials are defined as unique chemical conjugates of organic and/or inorganic materials [1]. That is, these are mixtures of two or more inorganic components, two or more organic components, or at least one of both types of components. The resulting material is not a simple mixture of its components but a synergistic material with properties and performance to develop applications with unique properties, which are determined by the interface of the components at the molecular or supramolecular level. Its functionality is associated with the improvement of physicochemical properties. For the electrochemical or biochemical properties through the optimization mainly of magnetic, electronic, optical, and thermal properties or a combination of them, since the mechanical properties are rather directing towards flexibility, which is not considered as a demand [1], see **Figure 1**. The inclusion of nano-sized materials has further expanded the extraordinary properties of hybrid materials thanks to the challenge of having greater options for multifunctional materials. In the last two decades, the development of multifunctional applications is receiving a lot of attention thanks to the chemical and physical properties of the materials, which are giving rise to developments of high adding value. These materials are classified as innovative advanced materials applicable to a huge diversity of applications including optics, electronics, sensors, ionics, energy conversion and storage, mechanics, membranes, protective coatings, catalysis, etc. [1].

IntechOpen

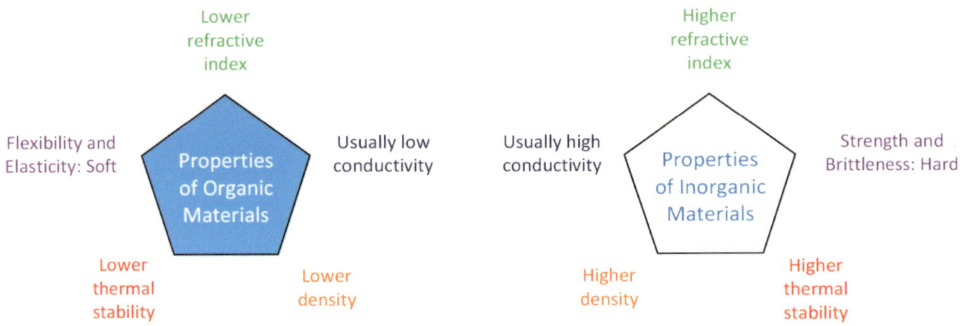

Figure 1.
Physical properties of the components of a hybrid material.

The unique versatility of these materials allows designing materials with tunable properties, with improved performance and properties to their long-established counterparts in the market. The diversity of organic and inorganic components that can be incorporated into these materials is from sizes of a few angstroms to thousands of angstroms, so these can be categorized among molecular species, in nano- and/or supramolecular sizes, or with extended structure. Some of the illustrative examples of hybrid materials are [1]:

- Donor-acceptor perovskites
- Intergrowth organic–inorganic perovskites
- Sol–gel silica modified with organic molecules
- Active organic molecules doped into conductive polymers
- Organically grafted inorganic phases
- Organically modified mesoporous materials
- Poly-oligo-silsesquixane-loaded polymers
- Active organic molecules intercalated into layered silicates, oxides, chalcogenides, and 2D materials
- Mixed organic–inorganic polymers
- Silica plus organic polymer
- Polymer-supported inorganic clusters or nanoparticles
- Polymers intercalated in layered silicates, oxides, chalcogenides, and 2D materials
- Polymer-clay composites
- Polymer-magnetic nanoparticle composites
- Biomineral-type composites

- Organically substituted polysiloxanes and polysilsesquioxanes
- Polymer-coated inorganic nanoparticles
- Silica-embedded bioactive species
- Fiber-reinforced nanocomposites, etc.

In the range of nanomaterials and supramolecular materials, there is a greater variety of possible cases, which leads to a wide continuous set, whose size ranges from molecules to solid-state materials. Moreover, the chemical nature of the components as well as the interaction between them leads to different possibilities of structure, degree of organization, and properties. In the design of this type of materials, it is transcendent to tune the nature, extent, and accessibility of internal surfaces [2–4]. Globally, the trend most used for the application of new materials is to predict and control their chemical and physical properties. In particular, the manipulation of atoms and molecules in nano-sized materials is related to nanochemistry and molecular engineering.

3. Classification

In accordance with the chemical origin of the interface or links established between the components in a hybrid material, these materials can be categorized as is shown in **Figure 2** [1, 2]. A first class of hybrid materials are those based on the synergy of the phases through weak chemical interactions based on Coulomb forces, London dispersion forces, hydrogen bonds, and dipole–dipole forces. A second class of hybrid materials are those based on the synergy of the phases through strong chemical bonds such as Lewis acid–base, covalent, or ionic-covalent bonds. The latter class of materials depends on the relative stability of the synergy between its components, since it determines the types of organic functionalization or the type of complex organic ligatures based on transition metal cations required to anchor the organic components to inorganic components.

In addition, hybrid materials can be classified as organic-in-inorganic (organic moieties used to modify inorganic materials) or inorganic-in-organic (inorganic constituents used to modify organic materials or matrices), as illustrated in **Figure 3** [3]. Hybrid materials based on the first approach can be subdivided into two types, namely, (1) inorganic materials modified by organic moieties and (2) colloidal polymers stabilized by organic moieties. On the other hand, inorganic materials are

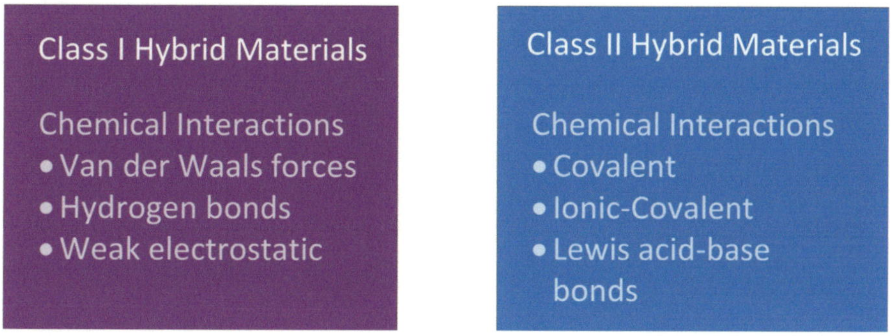

Figure 2.
Basic classification of hybrid materials according to the types of chemical interaction between their components.

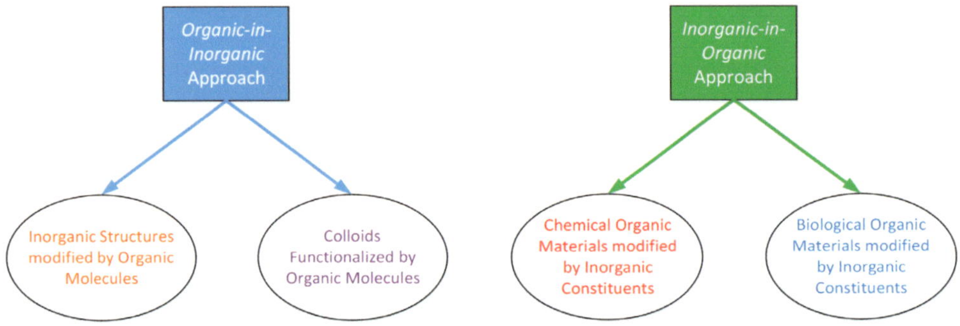

Figure 3.
Types of hybrid materials based on the adding of inorganic and organic components and vice versa.

modified via surface charge or functionalization with ligatures. Among the colloidal particles that can be used are nanoparticles, nanobars, nanotubes, nanostars, nanoflowers, etc., which must be electrostatically stabilized to be uniformly distributed throughout the material, preventing the formation of clusters. The incorporated inorganic constituents have small particle or structure morphologies, and these are made of clays, ceramics, minerals, metals, semiconductors, carbon-based nanomaterials, and two-dimensional materials. These materials are integrated into organic materials or matrices of either chemical or biological type. Chemical matrices can be polymers, monomers, synthetic molecules, etc., and the chemical materials derived from them are layers by layers (LbL), hydrogels, brushes, and copolymer blocks, both in the form of vehicles or coatings. The biological matrices used belong to one of the following groups: bacteria, microorganisms, molecules, polysaccharides, proteins, nucleic acids, carbohydrates, and lipids.

A more recent classification of hybrid materials is based on their functionality [1]. Three different types of materials can be identified: (1) structurally hybridized materials, (2) functionally hybridized materials, and (3) hybridized materials in their chemical bond.

4. Applications

Many applications have emerged by taking hybrid materials to the commercialization stage, and another huge amount is in its research and prototype phase to become emerging applications, as shown in **Figure 4** [4]. Among the numerous applications of organic–inorganic hybrid materials are:

- Gas barriers
- Packaging
- Sealants
- Hybrid pigments
- Decorative coatings
- Scratch-resistant coatings
- Anti-corrosion coatings
- Hair care products

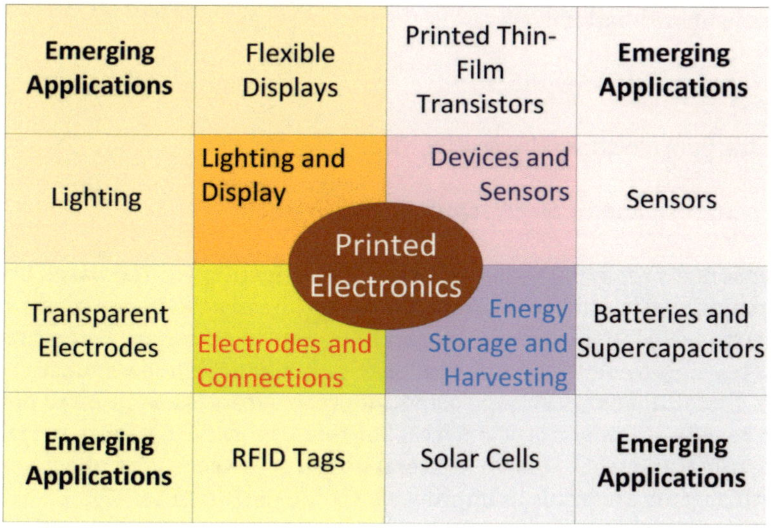

Figure 4.
Applications of hybrid nanomaterials to flexible electronics (adapted from [4]).

- Spin-on-glass materials
- Acoustic and thermal insulators
- Electrical insulators
- Smart textiles
- Flame retardants
- Green tires
- Automotive parts
- Dental products
- Controlled-release biocapsules
- Biocatalysts and/or photocatalysts
- Contrast agents for magnetic resonance imaging (MRI)
- Hybrid anti-cancer nanoparticles
- Biosensors
- Supercapacitors
- Fuel cells
- Solar cells
- Actuators
- Optical chemical sensors

- Flexible hybrid batteries
- Microlenses and waveguides
- Organic light-emitting diodes (OLEDs)
- Photochromic and/or electrochromic coatings, etc. [5]

Tag sensors for realizing radiofrequency identification (RFID) based on inkjet printing nanomaterials are easily stamped on textile, plastic, paper, glass, and metallic surfaces [6]. For example, by means of hybrid materials, using titania and silica, it is possible to develop templates onto polymer and/or glass substrates.

Electrochemical energy storage using supercapacitors is an option to power portable and/or wearable electronic devices. For this application, nanomaterials such as metal–organic frameworks (MOFs), metal nitrides, MXenes, and phosphorene are mixed with organic materials to improve electrode performance [7].

A huge variety of hybrid materials has been proposed for implementing electrodes for rechargeable batteries by means of inorganic polymers and materials such as graphene, carbon nanotubes, or their combination [8]. These storage devices are used to power portable applications.

Hybrid materials such as conductive polymers combined with nanostructured transition metal oxides, graphene, and/or carbon nanotubes are being used for the design of electrodes for solar cells [9].

5. Conclusions

Around the world, different research groups are continuously presenting new strategies, studies, and applications related to the technological development of novel hybrid materials. These materials promote continuous innovation in various technological sectors as presented in this chapter. Although there are numerous advances so far, the real possibilities can only be limited by the imagination of engineers and scientists around the world. The contributions of biologists, chemists, physicists, and materials scientists take advantage of both the integration and the miniaturization of electronic devices to develop emerging technologies in various areas of electronics as presented in the chapter. In this way, it can be affirmed that the area of hybrid nanomaterials is experiencing continuous growth as a topic of scientific research, incorporating more and more research groups worldwide thanks to the diversity of approaches that drive technological innovation.

Acknowledgements

The author appreciates the support of CONRICYT for access to the database to cite technical documents included in this chapter.

Thanks

The author wants to thank his wife and son for their support and time to edit this book. The author appreciates the support of Rebekah Pribetic working for Intech as Author Service Manager.

Author details

Rafael Vargas-Bernal
Higher Technological Institute of Irapuato, Carretera Irapuato-Silao Km,
Guanajuato, México

*Address all correspondence to: ravargas@itesi.edu.mx

References

[1] Gómez-Romero P, Sanchez C. Functional Hybrid Materials. Weinheim, Germany: Wiley-VCH; 2004. p. 434

[2] Kickelbick G. Hybrid Materials: Synthesis, Characterization and Applications. Germany Wiley-VCH: Weinheim; 2007. p. 517

[3] Saveleva MS, Eftekhari K, Abalymov A, Douglas TEL, Volodkin D, Parakhonskyiy V, et al. Hierarchy of hybrid materials – The place of inorganics-in-organics in it, their composition and applications. Frontiers in Chemistry. 2019;**7**:179. DOI: 10.3389/fchem.2019.00179

[4] Wu W. Inorganic nanomaterials for printed electronics: A review. Nanoscale. 2017;**9**(22):7342-7372. DOI: 10.1039/c7nr01604b

[5] Sanchez C, Bellevile P, Popall M, Nicole L. Applications of advanced hybrid organic-inorganic nanomaterials: From laboratory to market. Chemical Society Reviews. 2011;**40**(2):696-753. DOI: 10.1039/c0cs00136h

[6] Singh R, Singh E, Nalwa HS. Inkjet printed nanomaterials based flexible radio frequency identification (RFID) tag sensors for the Internet of nano things. RSC Advances. 2017;**7**(77):48597-48630. DOI: 10.1039/c7ra07191d

[7] Dubal DP, Chodankar NR, Km DH, Gomez-Romero P. Towards flexible solid-state supercapacitors for smart and wearable electronics. Chemical Society Reviews. 2018;**47**(6):2065-2129. DOI: 10.1039/c7cs00505a

[8] Peng HJ, Huang JQ, Zhang Q. A review of flexible lithium-sulfur and analogous alkali metal-chalcogen rechargeable batteries. Chemical Society Reviews. 2017;**46**(17):5237-5288. DOI: 10.1039/c7cs00139h

[9] Li L, Wu Z, Yuan S, Zhang XB. Advances and challenges for flexible energy storage and conversion devices and systems. Energy & Environmental Science. 2014;**7**(7):2101-2122. DOI: 10.1039/c4ee00318g

Chapter 2

Electronic Transport in Few-Layer Black Phosphorus

Gen Long, Xiaolong Chen, Shuigang Xu and Ning Wang

Abstract

Subjected to an adequately high magnetic field, Landau levels (LLs) form to alter the electronic transport behavior of a semiconductor. Especially in two-dimensional (2D) limit, quantum Hall effect sheds light on a variety of intrinsic properties of 2D electronic systems. With the raising quality of field effect transistors (FET) based on few-layer black phosphorus (BP), electronic transport in quantum limit (quantum transport) has been extensively studied in literatures. This chapter investigates the electronic transport in few-layer BP, especially in quantum limit. At the beginning of this chapter, a brief introduction to the background of LL, edge state, and quantum Hall effect will be delivered. We then examine the fabrication of high-quality FET based on BP and their electronic performances followed by exploring the magnetoresistances of these high-quality devices which reveal Shubnikov-de Haas (SdH) oscillations and quantum Hall effect in BP. Intrinsic parameters like effective mass, Landé g-factor, and so on are discussed based on quantum transport.

Keywords: black phosphorus, field effect transistor, mobility, Landau level, quantum hall effect

1. Introduction

Developing high-quality functional materials such as high-mobility semiconductors is an essential process to explore fundamental condensed matter research. The recent rediscovery of BP as a new member of 2D materials with theoretically predicted high mobility and widely tunable electronic bandgap makes it promising for various electronic devices and for probing interesting physical phenomena [1–4]. However, there are various obstacles to overcome in order to achieve the theoretically high carrier mobility in few-layer BP. Those obstacles include degradation of flake quality in atmosphere environment, electronic scatterings (against charge impurities, crystal defeats, and so on), high contact resistance, and so on. In devices without high enough mobility, new physical phenomena can be smeared out due to a large charge carrier scattering rate.

Since Landau level (LL) is crucial to understanding the physical phenomena in quantum transport process, we would like to deliver a brief idea about it here. For clarity, we will neglect the spin degeneracy at first and will come to it when necessary. Consider an electronic system without electronic interactions in an applied magnetic field along z direction and the magnetic field strength is B:

$$\boldsymbol{B} = \begin{pmatrix} 0 \\ 0 \\ B \end{pmatrix} \tag{1}$$

In Landau gauge (the wave function is gauge invariant),

$$\boldsymbol{B} = \nabla \times \boldsymbol{A} = \nabla \times \begin{pmatrix} 0 \\ Bx \\ 0 \end{pmatrix} \tag{2}$$

where $\boldsymbol{A}$ is the electromagnetic vector potential. The Hamiltonian of this system is then

$$H = \frac{1}{2m}(\boldsymbol{P} - e\boldsymbol{A})^2 = \frac{p_x^2}{2m} + \frac{1}{2m}\left(p_y - eBx\right) \tag{3}$$

where e stands for electron charge, $\boldsymbol{P}$ is the canonical momentum, and m is the mass of charge carriers. Solving this Hamiltonian we get,

$$\begin{aligned} E_n &= \hbar\omega(n + 1/2) \\ \omega &= \frac{e\boldsymbol{B}}{m} \end{aligned} \tag{4}$$

where ω is the cyclotron frequency and $\hbar$ is the reduced Planck constant. Clearly, the eigen energy of electrons are separated and equally spaced. The energy gap between two adjacent LLs is $\hbar\omega$. Furthermore, by studying the electronic systems in finite regime, the degeneracy (g_L) of LL per unit area is derived to be,

$$g_L = \frac{eB}{2\pi\hbar} \tag{5}$$

Now we take the spin of electron into consideration. Besides the formation of LLs, the energy of electron is also split by Zeeman effect $E_Z = g\mu_B B$, where $\mu_B = e\hbar/2m_0$ (m_0 is free electron mass) is the Bohr magneton and g is the Landé g-factor. Interestingly, the Zeeman splitting coincides precisely with the energy gap between LLs $E_C = \hbar\omega = e\hbar B/m$ [5], for free electrons ($g = 2, m = m_0$). It means that spin-up electrons in n_{th} LL exhibit the same energy with spin down ones in $(n+1)_{th}$ LLs. However, this coincidence usually does not happen in real material systems because of the deviations of effective mass from m_0 and g-factor from 2. Nevertheless, this coincidence can be realized by tilting the magnetic field to adjust the ratio between E_C and E_Z which would be discussed hereafter.

We conclude now that under applied magnetic field, electrons cycle in quantized traces with quantized energy in one direction. However, when these traces intersect the edge of the electron system, these electrons cannot complete the full cyclotron. Instead, they bounce back and enter another adjacent traces because the cycling direction is fixed. As a result, electrons near the edge are in a skipping motion along one-dimensional boundary moving in a single direction, which is referred to as chiral edge states. Due to the absence of scattering, the edge states carry electronic current with no voltage drop, i.e., $R_{xx} = V_{xx}/I_{channel} = 0$. Electrons near different edges move in different directions to ensure that the net current, in the absence of electric field, is zero.

When an in-plane electric field is applied, electrons are accelerated along one direction. As a result, the balance between the occupation of edge states in two opposite sides is broken, i.e., more electrons are filled in the edge states in one boundary. As we have discussed before, the energy of electrons is quantized. Consequently, the Hall conductance $\sigma_H = \frac{I_{channel}}{V_H}$ is quantized. Analysis into more details leads to $\sigma_H = \nu\frac{e^2}{h}$, where $\nu = 1, 2, 3...$ is integer.

Alternatively, quantum Hall effect can be understood based on Berry phase [6]. Consider an adiabatic system in which the eigenstate evolves with external parameters slowly. When the external parameters make up a loop and come back to itself, the eigenstate should also come back to itself (let us rule out the possibility of degeneracy here) but with a different phase, because the eigen energy of eigenstates with different phases is invariant. This phase is the so-called Berry phase. This Berry phase manifests itself in an electronic system through weak localization effect, A-B effect, and so on. All these effects come from the dynamics of electrons. For Bloch electron under the perturbation of a weak electric field $\boldsymbol{E}$, its velocity $\boldsymbol{v}$ with a given wave vector $\boldsymbol{k}$ becomes

$$\boldsymbol{v}(\boldsymbol{k}) = \frac{\partial \varepsilon(\boldsymbol{k})}{\hbar \partial \boldsymbol{k}} - \frac{e}{\hbar}\boldsymbol{E} \times \boldsymbol{\Omega}(\boldsymbol{k}) \tag{6}$$

where ε is the energy of electron and $\boldsymbol{\Omega}$ is the Berry curvature. The first term is the result of band dispersion. The second term which is absent in classic Bloch theorem is more interesting. It describes a velocity perpendicular to the applied electric field which will give rise to the Hall effect without magnetic field. With this transverse contribution to velocity, the Hall conductivity is given by

$$\sigma_{xy} = \frac{e^2}{\hbar}\int_{BZ} \frac{d^2k}{(2\pi)^2}\Omega_{k_x k_y} \tag{7}$$

where the integration is over the entire Brillouin zone and the result of this integration is the Chern number ν of the Brillouin zone ($\nu = 0$ when external magnetic field is zero). Therefore, the Hall conductivity is

$$\sigma_{xy} = \nu\frac{e^2}{h} \tag{8}$$

This reaches the same conclusion with previous analysis, i.e., the Hall conductivity is quantized.

In this chapter, we would introduce the detailed work on BP from the characterization of the electronic transport process and figure out the main limitation to high-quality devices, followed by efforts to conquer the impediment. Once devices with high enough quality can be fabricated, we will go to the electronic transport in quantum limit (cryogenic temperature and high magnetic field). Realization of quantum transport process in BP allows people to determine some critical parameters (effective mass, Landé g-factor, and so on) which are crucial for seeking engineering applications of this material.

2. High-quality FET based on BP

In 2014, several groups first demonstrated the FET based on few-layer BP (**Figure 1a**) [1, 3, 7, 8], giving rise to the renaissance of BP. As a semiconducting layered material, BP has a high carrier mobility at room temperature, with the value up to 1000 $cm^2V^{-1}s^{-1}$ at the first demonstration of its FET, owing to its relatively

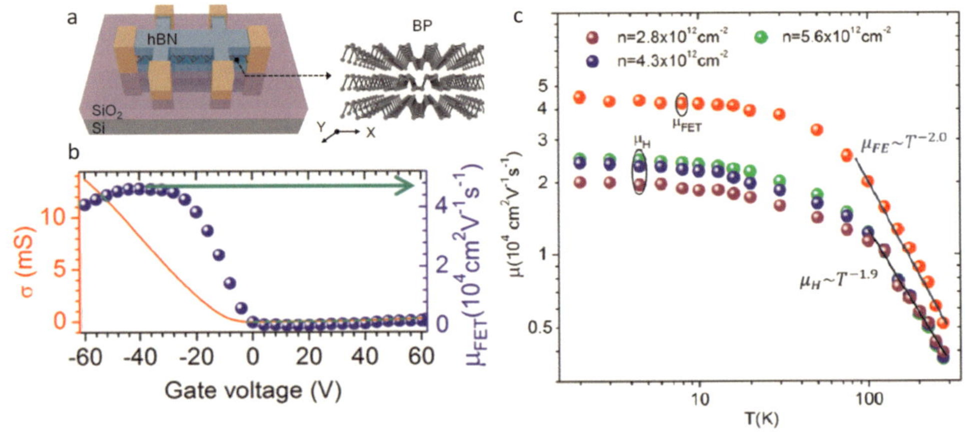

Figure 1.
High-mobility BP FET. (a) Schematic of FET device based on encapsulated BP. The right panel shows the crystal structure of BP. (b) Transfer curves of high-mobility BP devices at cryogenic temperature accompanied with the corresponding FET mobility. (c) Temperature dependence of FET mobility and hall mobility. Panel (a) is reproduced with permission from NPG [12]. Panels (b) and (c) are reproduced with permission from ACS [10].

weak electron-phonon interaction [3]. Meanwhile, the FET's current on-off ratio exceeds 10^5, which indicates the promising electronic application of BP-based devices [2, 3, 9]. Up to now, the record room temperature hole mobility achieved in experiments arrives 5200 $cm^2V^{-1}s^{-1}$, almost approaching the theoretical phonon scattering limitation ranging between 4800 and 6400 $cm^2V^{-1}s^{-1}$ for few-layer BP [10]. The high mobility and desirable current saturation of BP at room temperature offer advantages for high-frequency electronic and optoelectronic applications [11]. At cryogenic temperatures, the FET hole mobility can be even up to 45,000 $cm^2V^{-1}s^{-1}$ as shown in **Figure 1b** and **c** [10], the highest value among the presently reported 2D semiconductors. The ultrahigh mobility in BP at low temperature makes it as a unique platform for studying the quantum transport in 2D semiconductors.

As the most famous elemental layered material beyond graphene, BP provides the needed bandgap for FET applications, especially in infrared region. Unlike semiconducting transition metal dichalcogenides (TMDCs), the bandgap in BP is direct for all number of layers, which makes this material particularly promising for optoelectronic applications. Moreover, the bandgap of BP is widely tunable achieved by layer number, electric field, strain, and alloying. Combined with its strong anisotropic, BP may allow for the exploration of new exotic phenomena and multifunction devices.

2.1 Realization of air-stable devices based on BP

Although bulk BP is considered as the most stable allotrope of phosphorus, it is reported that few-layer BP is unstable in ambient conditions. Significant surface roughening over the time after exfoliation can be observed by AFM [8]. The investigation of water condensation on BP flakes reveals that BP is very hygroscopic and tends to uptake moisture from air [13]. Long-time exposure to air can even etch BP away. The in situ Raman and transmission electron spectroscopy studies of the degradation of BP show that the degradation in air mainly arises from the photo-assisted oxidation reaction with oxygen dissolved in the adsorbed water on its surface [14]. Theoretical work revealed that BP presents a strong dipolar moment out of plane which makes it very hydrophilic [15].

The environmental instability of BP flakes makes it a challenge to fabricate high performance BP-based devices, since most experiments reported on BP relied on mechanically exfoliated flakes. To prevent the degradation, several kinds of passivation were developed. Atomic layer-deposited AlO_x overlayers effectively suppress ambient degradation, preserving the intrinsic high carrier mobility and on-off ratios in BP FETs [16]. Alternatively, encapsulating BP by a polymer superstrate, such as PMMA, can also suppress the oxidation [8, 17]. Moreover, it was demonstrated that hexagonal boron nitride (*h*BN) can be effectively used for passivation of BP [9, 18–21]. The devices fabricated by *h*BN-encapsulated BP showed air-stable performance and hysteresis-free transport characteristics in ambient conditions, without observable decrease in carrier mobility or on-off ratio after few weeks' exposure in air [9, 18]. Furthermore, *h*BN is an insulating layered material, providing atomically thin clean surface and dielectric environment. Before the rediscovery of BP, it was already reported that *h*BN can significantly improve the mobility of graphene-based devices [22, 23]. Researchers found that by applying the *h*BN encapsulation technique in graphene to BP (**Figure 2**), the mobility of BP-based FETs can increase to several 1000 $cm^2V^{-1}s^{-1}$ [9, 24], later even up to 45,000 $cm^2V^{-1}s^{-1}$ [10] at cryogenic temperatures.

2.2 Ohmic contact and ambipolar transport in BP

As that in other semiconducting devices, the electrical performance of devices based on BP is significantly affected by the electrical contacts. Low contact resistance is critical to achieving high mobility, high on-off ratio, and large photo response in BP. The main issue in BP-based FET is the existence of large Schottky barrier in the contacts, which limits the current injection and its potential for applications. Several kinds of contact engineering have been identified toward high-quality electrical contacts. By choosing various metals with different work functions to match BP, different performance of BP transistors can be achieved [26, 27]. The aluminum-contacted BP displays ambipolar characteristic with nearly symmetric electron and hole mobility of 950 $cm^2V^{-1}s^{-1}$ in 13 nm flake and unipolar *n*-type

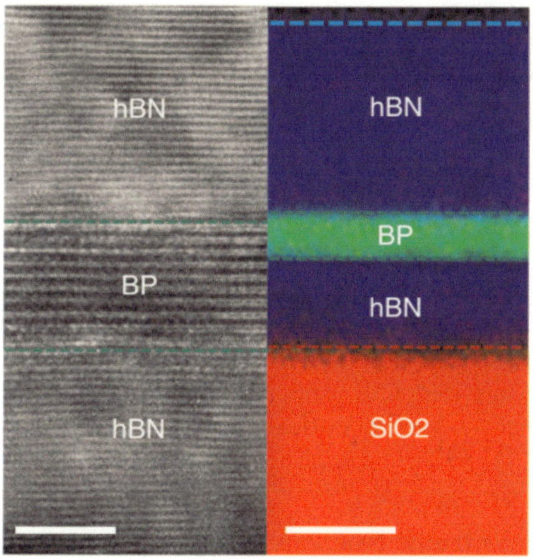

Figure 2.
The cross-section HR-TEM image of the hBN/BP/hBN *structure. The right panel shows the element analysis for nitrogen (blue), phosphorus (green), and oxygen (red). Scale bars: 4/10 nm (left/right). This figure is reproduced with permission from NPG [25].*

behavior in 3 nm flake, while the palladium-contacted BP shows *p*-type dominated transport behaviour in thick flakes. Besides traditional metallic contacts, graphene can also be used as a contact medium to BP. Graphene has an atomically flat surface, and its work function can be tuned by gate voltages, which leads to the elimination of Schottky barrier height using graphene electrodes [18]. In addition, to reduce the contact resistance, monolayer *h*BN can be inserted between metal films and BP by van der Waals transfer technique [20]. The monolayer *h*BN acts as a tunnel barrier with relatively high tunnel conductivity. If bilayer *h*BN is used as the tunnel barrier, it results in a higher contact resistance with a factor of 10 than that in monolayer *h*BN. In *h*BN-encapsulated BP devices, selective area etching technique was used to open the window to make the contact [9]. By using appropriate etching recipe, the top *h*BN can be quickly etched, while BP layer still survives, ensuring the metals directly contact with the plasma-treated BP. Low contact resistance down to several $k\Omega \cdot \mu m$ can be achieved even at low temperature. Surprisingly, this selective area etching technique can be effectively applied to other 2D semiconductor such as TMDCs with high performance [28].

Few-layer BP has a direct bandgap with the size around 0.3 eV. The small bandgap of few-layer BP facilitates the injections of both electron and hole, resulting in ambipolar characteristic being easily observed in BP-based transistors. The ambipolar transport in BP can exploit both *n*-type and *p*-type in a single transistor, with the mobility of both electron and hole at a high value of several thousand $cm^2V^{-1}s^{-1}$ [29], which provides the promising applications in complementary metal-oxide semiconductors and 2D material-based memory devices [30]. The possibility of manipulation in a single ambipolar transistor without external doping or ion implantation process simplifies the BP-based nanotechnology in feature's industrial application. The ambipolar functionality and high mobility of BP help the realization of flexible ambipolar inverter, frequency doubler, inverting and noninverting analogy amplifiers, and amplitude-modulated demodulator [31]. Furthermore, the ambipolar operation of BP can be locally controlled by electrostatic gating to form gate-defined PN junction [32]. Under illumination, these PN junctions show strong photocurrent due to photovoltaic effect, attractive for energy harvesting in the near-infrared.

2.3 Strong transport anisotropy in BP

BP has a puckered honeycomb lattice, yielding strong in-plane anisotropy. The unusual anisotropic structure of BP results in its strong in-plane anisotropic electrical and optical properties. BP-based devices with electrode probes fabricated at various angles have been used to probe the electrical conductivity and carrier mobility along different direction (**Figure 3**) [1, 7]. Higher hole mobility and conductivity were found along light effective mass directions (x-direction) [1, 9]. The experimental results verified the theoretical calculations, which predicted that the effective mass in few-layer BP along x and y directions are $m_x \sim 0.14\, m_0$ and $m_y \sim 0.89\, m_0$, respectively [2]. Not only electrical properties but also optical properties of BP show strong in-plane anisotropy, such as optical absorption and polarized Raman spectrum in few-layer BP [1] and photoluminescence in monolayer BP [33].

2.4 Widely tunable electronic bandgap

Due to the strong layer interaction and quantum confinement of the charge carriers in the out-of-plane direction, BP shows stronger thickness-dependent bandgap compared with other 2D semiconductors, such as TMDCs. When the

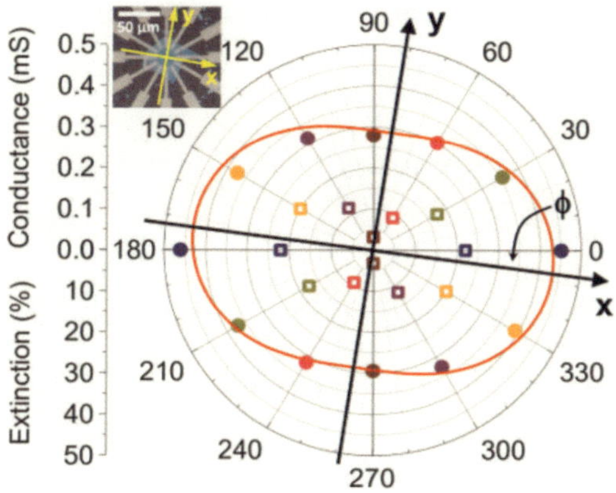

Figure 3.
Anisotropic transport of BP. This figure is reproduced with permission from NPG [1].

thickness of BP decreases from bulk to monolayer, its direct bandgap was predicted to increase roughly from 0.3 to 2.0 eV [34]. This prediction has been experimentally confirmed by the measurements of optical absorption [35]. The optical bandgaps of monolayer, bilayer, trilayer, and bulk BPs were determined to be 1.73, 1.15, 0.83, and 0.35 eV, respectively. The bandgap of few-layer BP is also widely tunable by electric fields. By using an in situ potassium doping technique, a vertical electric field from dopants modulates the bandgap of few-layer BP and even tunes the system to an anisotropic Dirac semimetal state, owing to the giant Stark effect [36]. For practical applications, the dynamic tuning of bandgap can be realized by the dual-gate FET configuration in BP-based devices [37]. For example, the transport measurement in a 10-nm-thick BP flake shows that its bandgap can be continuously tuned from 0.3 eV to below 0.05 eV, using a moderate displacement field of 1.1 Vnm^{-1}. Local-strain engineering was demonstrated to be another way to control the bandgap in BP. When BP is subjected to a periodic strain, remarkable shift of optical absorption edge up to 0.7 eV was detected [38]. When BP-based FETs fabricated flexible substrates, the applied mechanical strain continuously modulates its bandgap, significantly altering the density of thermally activated carriers. As a result, large piezo-resistive effects were observed at room temperature [39]. Another way to tune the bandgap of BP includes alloying. Successful synthesis of layered black arsenic-phosphorus with tunable compositions allows the bandgap of alloyed BP to be changed from 0.3 to 0.15 eV [40], making black arsenic-phosphorus as a promising candidate for long-wavelength infrared photodetectors [41, 42].

3. SdH oscillation and quantum hall effect in BP

Upon subjection to a magnetic field, LLs form in an electronic system. However, high scattering rate of charge carriers leads to a broadening of LLs. The LLs manifest itself only when this broadening is lower than the energy gap between two adjacent LLs. The achievement of high-quality devices (high carrier mobility and low scattering rate) suppresses this broadening and finally allows the observation of LL with laboratory reachable magnetic fields. By the end of 2014, Li et al. and Nathaniel et al. reported the first observation of Shubnikov-de Haas (SdH) oscillations in BP hole system [3, 24] (**Figure 4a**). Later on Li et al. observed the quantum Hall effect in the same system (**Figure 4b**) [43]. **Figure 4a** shows ΔR_{xx} (a smooth background subtracted from R_{xx}) and striking patterns of SdH oscillations,

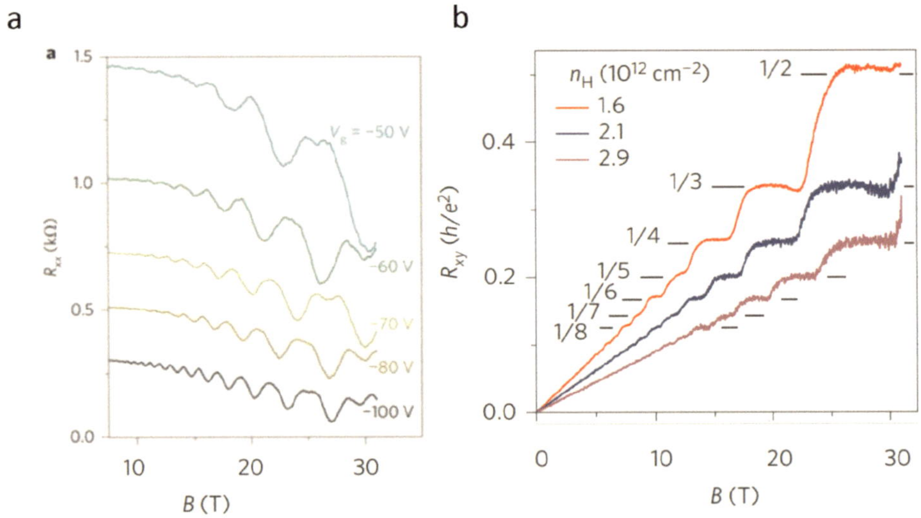

Figure 4.
Quantum transport in BP. (a) R_{xx} versus V_g and magnetic field B. (b) Hall resistance $R_{xy} = V_{xy}/I$ variation with magnetic field at cryogenic temperature. The quantized plateaus demonstrate the quantum Hall effect in BP two-dimensional hole gas. This figure is reproduced with permission from NPG [43, 44].

appearing along straight lines, are clearly resolved. When the Fermi level of the 2D hole system is located at the center of certain LLs (at the middle of two adjacent LLs), the system exhibits high resistance (low resistance). Hence the oscillations are employed as a powerful tool to monitor the evolution of Fermi energy with gate voltage considering that the energy gap between two adjacent LLs is $\hbar\omega$.

More quantitatively, according to the Lifshitz-Kosevich model, the temperature dependence of oscillation amplitude follows [45].

$$\Delta R \propto \frac{\lambda(T)}{sinh\ (\lambda(T))} \tag{9}$$

where $\lambda(T) = 2\pi^2 k_B T m^* / \hbar e B$ is the thermal damping factor, k_B stands for Boltzmann constant, $\hbar$ denotes the reduced Plank constant, and m^* is the effective cyclotron mass. **Figure 5a** shows the oscillation component ΔR_{xx} evolves with magnetic field at few different temperatures. The obtained temperature dependence of the oscillation amplitudes accompanied with the fitting result of thermal reduction factor is shown in **Figure 5b** [44]. **Figure 5c** displays the extracted effective cyclotron mass from the fitting result [44]. Similar values of effective cyclotron mass have been reported by different groups [9, 10, 24, 27].

As device quality further improved, Zeeman splitting can be resolved in laboratory-accessible magnetic field [10], and alternative SdH oscillation amplitudes were observed. By reproducing the oscillation component with high-order spin-resolved LK formula

$$\Delta R_{xx} = 2R_0 \sum_{r,\uparrow,\downarrow} \frac{r\lambda(T)}{\sinh\left(\lambda(T)\right)} \exp\left(-r\frac{\pi}{\omega\tau_{\uparrow,\downarrow}}\right) \cos\left(r\phi_{\uparrow,\downarrow}\right) \tag{10}$$

where ω is the cyclotron frequency, $\lambda(T)$ stands for the thermal factor, $\phi_{\uparrow,\downarrow}$ is the corrected Berry phase taking Zeeman splitting into consideration, and $\tau_{\uparrow,\downarrow}$ is the spin-resolved quantum scattering times. The extracted spin-dependent quantum scattering time at different temperatures and gate voltages are presented in

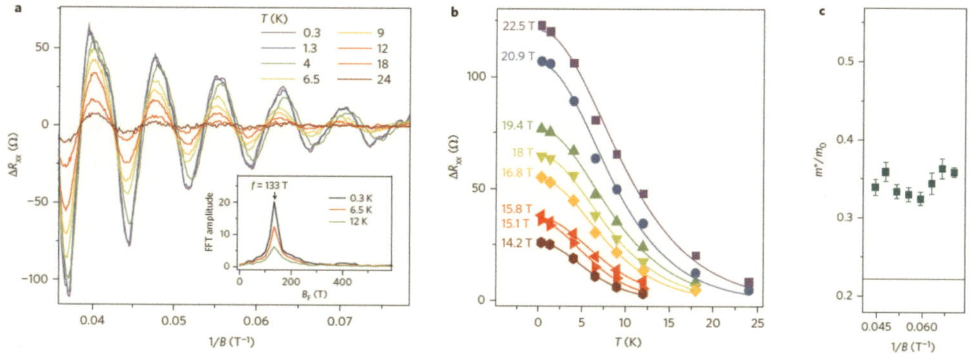

Figure 5.
Effective mass of holes in BP. (a) ΔR_{xx} as a function of 1/B at different temperatures. The inset shows the FFT result at selected temperatures. (b) Evolution of oscillation amplitudes with temperatures at different magnetic fields. The solid lines represent the fitting result of temperature reduction factor. (c) Effective mass of holes obtained from the fitting results shown in panel b. This figure is reproduced with permission from NPG [44].

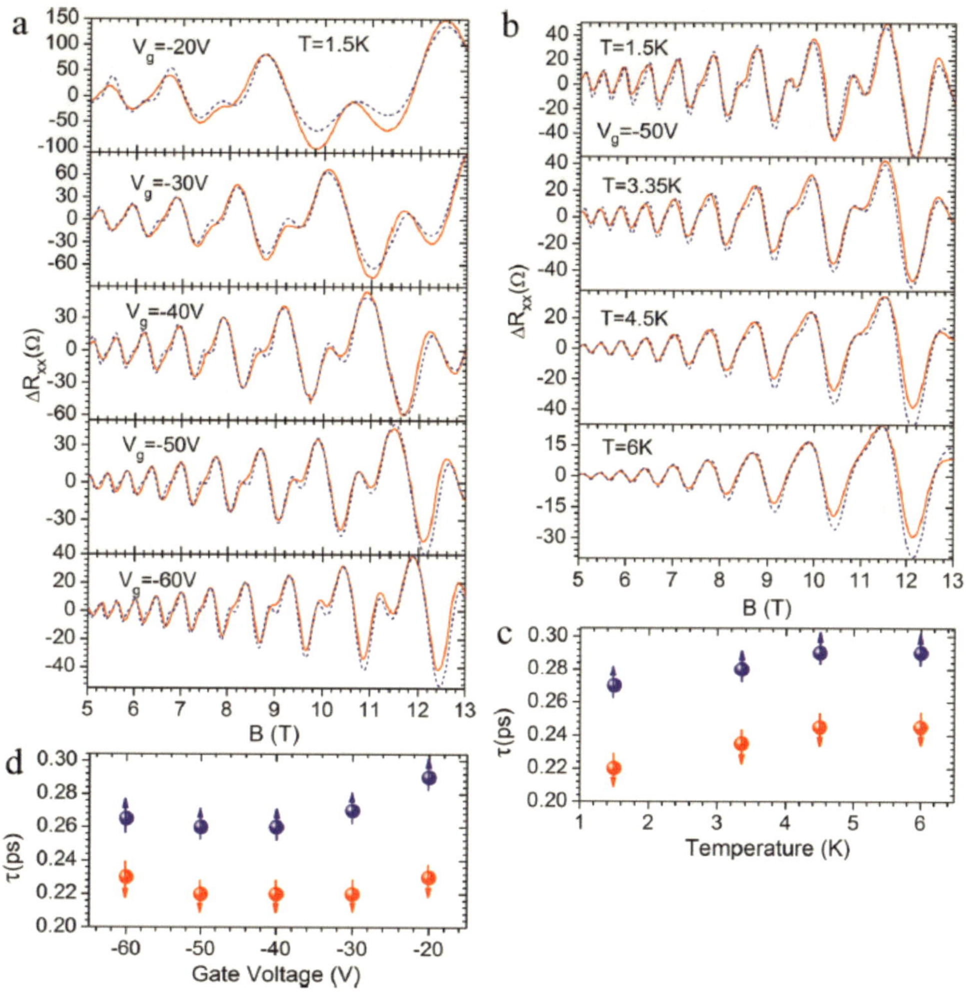

Figure 6.
Spin-selective quantum scattering process (a, b) Oscillation component ΔR_{xx} at selected gate voltages (panel a) and temperatures (panel b). The blue dashed lines present the fitting results of Eq. (10). (c, d) Quantum scattering times for up- and down-spin orientations obtained from the fitting results. This figure is reproduced with permission from ACS [10].

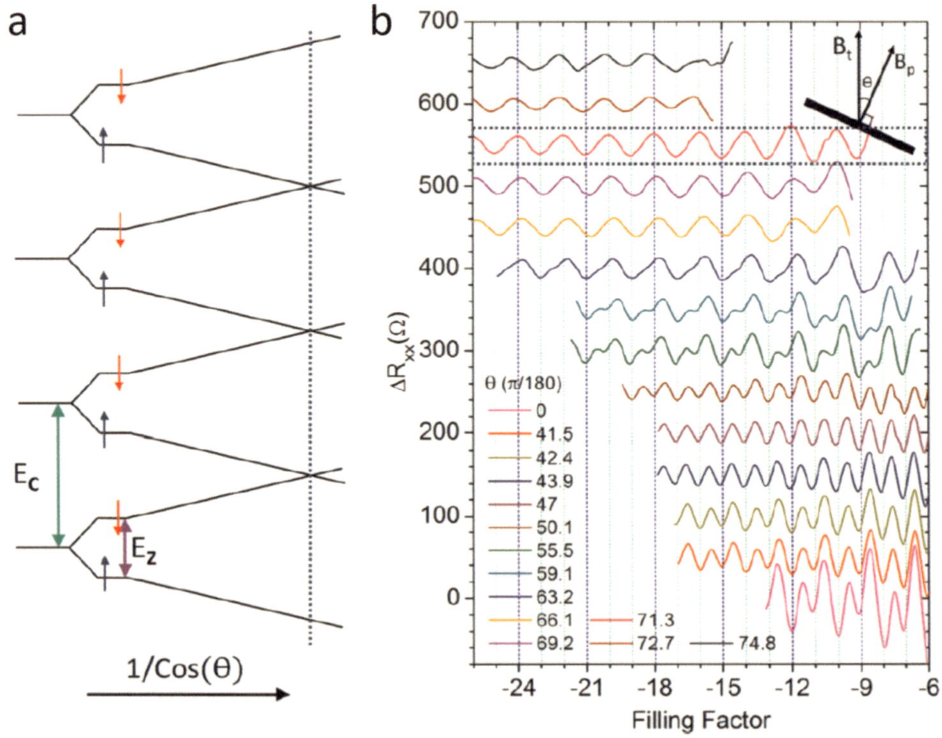

Figure 7.
Coincidence of BP LLs (a) Schematic of spin-resolved LLs evolve with increasing tilting angle at fixed perpendicular magnetic field. The vertical dashed line indicates the coincidence angle θ_c. (b) The evolution of SdH oscillation component with increasing tilting angles. This figure is reproduced with permission from ACS [10].

Figure 6c and **d**, respectively. The charge carriers in BP two-dimensional hole gas (2DHG) show the spin-selective scattering behavior on a phenomenological level. Similar scattering process was also reported in GaAs [46] and ZnO [47] systems. Recently, an analogy called valley-selective scattering process was observed in MoS_2 [48]. The mechanism behind this spin-selective scattering process remains to be further addressed.

The resolvability of Zeeman splitting makes it possible to determine the Landé g-factor of BP electronic systems via the coincidence method taking advantage that the Zeeman energy $E_Z = g\mu_B B_{total}$ depends on the total magnetic field, while the cyclotron energy $E_C = \hbar eB_\perp/m^*$ is determined by the field component $B_\perp$ perpendicular to the 2DHG plane. When the E_Z is a multiple integer of E_C, i.e., $E_Z = iE_C$ (i is an integer number), the spin-resolved LLs overlap. Hence $\chi_s = gm^* = 2i\cos\theta_c$ is valid at coincidence angle θ_c as indicated in **Figure 7a** [10, 49]. **Figure 7b** presents the measurement of coincidence angle, and the extracted spin susceptibility χ_s is $\sim$ 0.64 which leads to a Landé g-factor of $g = 2.47$ [10]. Later on, the same method was applied to electron-doped BP to extract a spin susceptibility of $\sim$ 1.1 and Landé g-factor of $\sim$ 2.8 [29, 50].

Making use of a magnetic field up to 45 T, fractional quantum state at filling factor $\nu = -\frac{4}{3}$ accompanied with quantum feature at -0.56 ± 0.1 was observed in 2018 [51] making BP the second exfoliated van der Waals crystals to exhibit fractional quantum Hall effect after graphene.

When the electronic band of a semiconductor is occupied by induced electrons (e.g., electrons induced by gate voltage), it will screen the electronic field hence prohibiting the wave function of electrons from penetrating the topmost few layers [52–54].

While when the band is empty, it will not screen the electronic field, and the wave function of electrons penetrates all the way to the opposite surface. Based on this screening effect, wide quantum wells have been formed in few-layer BP which can be switched between double layers of charge carriers and single layer of charge carriers [55]. The switch in one single device from double quantum wells and single quantum well has never been realized in other systems. Besides, one can tune the distance between the double layers of charge carriers by selecting *h*BN flakes with different thicknesses. The realization of tunable quantum wells paves the way for using 2D materials as wide quantum wells for investigating phenomena, such as LL hybridization, inter-well Coulomb interactions, or multicomponent quantum Hall ferromagnetism in this highly anisotropic system.

4. Conclusions

In this chapter, we mainly focus on the development of high-quality FET and the electronic transport in quantum limitations based on BP. Although few-layer BP degrades in the atmosphere, encapsulation with BN flakes was proven to be a reliable strategy to realize air-stable devices based on BP. The realization of Ohmic contact through selective etching technique lays the foundation for high-quality devices. Thanks to the rather small electronic bandgaps for few-layer BP flakes, by aligning the work function of contact metals with the band edge of BP, ambipolar-conducting channels were realized in single field effect devices. In devices with high carrier mobility, the scattering rate of charge carriers is low leading to limited broadening of LLs. As a consequence, the LLs become resolvable in laboratory reachable magnetic field at cryogenic temperature. Based on the SdH oscillations and quantum Hall effect, a spin-selective quantum scattering process was established in BP 2DHG. Intrinsic parameters such as effective mass and Landé g-factor were measured. The realization of high-quality air-stable BP devices promises its potential in next-generation electronic applications.

Acknowledgements

The authors acknowledge the financial support from the Research Grants Council of Hong Kong (Project No. FB417- UoM-HKUST and C7036-17 W).

Conflict of interest

The authors declare no conflict of interest.

Author details

Gen Long[1,2]†, Xiaolong Chen[1,3]†, Shuigang Xu[4]† and Ning Wang[1]*

1 The Hong Kong University of Science and Technology, Hong Kong, China

2 University of Geneva, Geneva, Switzerland

3 Southern University of Science and Technology, Shenzhen, China

4 University of Manchester, Manchester, UK

*Address all correspondence to: phwang@ust.hk

† These authors contributed equally.

References

[1] Xia F, Wang H, Jia Y. Rediscovering black phosphorus as an anisotropic layered material for optoelectronics and electronics. Nature Communications. 2014;**5**:4458

[2] Qiao J, Kong X, Hu Z-X, Yang F, Ji W. High-mobility transport anisotropy and linear dichroism in few-layer black phosphorus. Nature Communications. 2014;**5**:4475

[3] Li L, Yu Y, Ye GJ, Ge Q, Ou X, Wu H, et al. Black phosphorus field-effect transistors. Nature Nanotechnology. 2014;**9**(5):372

[4] Du Y, Liu H, Deng Y, Ye PD. Device perspective for black phosphorus field-effect transistors: Contact resistance, ambipolar behavior, and scaling. ACS Nano. 2014;**8**(10):10035-10042

[5] Sondhi SL, Karlhede A, Kivelson SA, Rezayi EH. Skyrmions and the crossover from the integer to fractional quantum hall effect at small zeeman energies. Physical Review B. 1993;**47**(24):16419

[6] Xiao D, Chang M-C, Niu Q. Berry phase effects on electronic properties. Reviews of Modern Physics. 2010;**82**(3): 1959

[7] Liu H, Neal AT, Zhu Z, Luo Z, Xu X, Tománek D, et al. Phosphorene: An unexplored 2D semiconductor with a high hole mobility. ACS Nano. 2014; **8**(4):4033-4041

[8] Koenig SP, Doganov RA, Schmidt H, Castro Neto AH, Özyilmaz B. Electric field effect in ultrathin black phosphorus. Applied Physics Letters. 2014;**104**(10):103106

[9] Chen X, Yingying W, Zefei W, Han Y, Shuigang X, Wang L, et al. High-quality sandwiched black phosphorus heterostructure and its quantum oscillations. Nature Communications. 2015;**6**:7315

[10] Long G, Maryenko D, Shen J, Shuigang X, Hou J, Zefei W, et al. Achieving ultrahigh carrier mobility in two-dimensional hole gas of black phosphorus. Nano Letters. 2016;**16**(12): 7768-7773

[11] Wang H, Wang X, Xia F, Wang L, Jiang H, Xia Q, et al. Black phosphorus radio-frequency transistors. Nano Letters. 2014;**14**(11):6424-6429

[12] Chen X, Chen C, Levi A, Houben L, Deng B, Yuan S, et al. Large-velocity saturation in thin-film black phosphorus transistors. ACS Nano. 2018;**12**(5): 5003-5010

[13] Castellanos-Gomez A, Vicarelli L, Prada E, Island JO, Narasimha-Acharya KL, Blanter SI, et al. Isolation and characterization of few-layer black phosphorus. 2D Materials. 2014;**1**(2): 025001

[14] Favron A, Gaufrès E, Fossard F, Phaneuf-L'Heureux A-L, Tang NYW, Lévesque PL, et al. Photooxidation and quantum confinement effects in exfoliated black phosphorus. Nature Materials. 2015;**14**(8):826

[15] Du Y, Ouyang C, Shi S, Lei M. Ab initio studies on atomic and electronic structures of black phosphorus. Journal of Applied Physics. 2010;**107**(9):093718

[16] Wood JD, Wells SA, Jariwala D, Chen K-S, Cho EK, Sangwan VK, et al. Effective passivation of exfoliated black phosphorus transistors against ambient degradation. Nano Letters. 2014;**14**(12): 6964-6970

[17] Tayari V, Hemsworth N, Fakih I, Favron A, Gaufrès E, Gervais G, et al. Two-dimensional magnetotransport in a black phosphorus naked quantum well. Nature Communications. 2015;**6**:7702

[18] Avsar A, Vera-Marun IJ, Tan JY, Watanabe K, Taniguchi T, Castro

Neto AH, et al. Air-stable transport in graphene-contacted, fully encapsulated ultrathin black phosphorus-based field-effect transistors. ACS Nano. 2015;**9**(4): 4138-4145

[19] Doganov RA, O'Farrell ECT, Koenig SP, Yeo Y, Ziletti A, Carvalho A, et al. Transport properties of pristine few-layer black phosphorus by van der waals passivation in an inert atmosphere. Nature Communications. 2015;**6**:6647

[20] Cao Y, Mishchenko A, Yu GL, Khestanova E, Rooney AP, Prestat E, et al. Quality heterostructures from two-dimensional crystals unstable in air by their assembly in inert atmosphere. Nano Letters. 2015;**15**(8):4914-4921

[21] Long G, Xu S, Cai X, Wu Z, Han T, Lin J, et al. Gate-tunable strong-weak localization transition in few-layer black phosphorus. Nanotechnology. 2017; **29**(3):5204

[22] Dean CR, Young AF, Meric I, Lee C, Wang L, Sorgenfrei S, et al. Boron nitride substrates for high-quality graphene electronics. Nature Nanotechnology. 2010;**5**(10):722

[23] Wang L, Meric I, Huang PY, Gao Q, Gao Y, Tran H, et al. One-dimensional electrical contact to a two-dimensional material. Science. 2013;**342**(6158): 614-617

[24] Gillgren N, Wickramaratne D, Shi Y, Espiritu T, Yang J, Hu J, et al. Gate tunable quantum oscillations in air-stable and high mobility few-layer phosphorene heterostructures. 2D Materials;**2**(1):011001, 2015

[25] Chen X, Lu X, Deng B, Sinai O, Shao Y, Li C, et al. Widely tunable black phosphorus mid-infrared photodetector. Nature Communications. 2017;**8**(1):1672

[26] Perello DJ, Chae SH, Song S, Lee YH. High-performance n-type black phosphorus transistors with type control via thickness and contact-metal engineering. Nature Communications. 2015;**6**:7809

[27] Long G, Xu S, Shen J, Hou J, Wu Z, Han T, et al. Type-controlled nanodevices based on encapsulated few-layer black phosphorus for quantum transport. 2D Materials. 2016;**3**(3): 031001

[28] Xu S, Wu Z, Lu H, Han Y, Long G, Chen X, et al. Universal low-temperature ohmic contacts for quantum transport in transition metal dichalcogenides. 2D Materials. 2016; **3**(2):021007

[29] Long G, Maryenko D, Pezzini S, Xu S, Wu Z, Han T, et al. Ambipolar quantum transport in few-layer black phosphorus. Physical Review B. 2017; **96**(15):155448

[30] He T, Deng B, Chin ML, Yan X, Jiang H, Han S-J, et al. A dynamically reconfigurable ambipolar black phosphorus memory device. ACS Nano. 2016;**10**(11):10428-10435

[31] Zhu W, Yogeesh MN, Yang S, Aldave SH, Kim J-S, Sonde S, et al. Flexible black phosphorus ambipolar transistors, circuits and AM demodulator. Nano Letters. 2015;**15**(3): 1883-1890

[32] Buscema M, Groenendijk DJ, Steele GA, Van Der Zant HS, Castellanos-Gomez A. Photovoltaic effect in few-layer black phosphorus pn junctions defined by local electrostatic gating. Nature Communications. 2014;**5**: 4651

[33] Wang X, Jones AM, Seyler KL, Tran V, Jia Y, Zhao H, et al. Highly anisotropic and robust excitons in monolayer black phosphorus. Nature Nanotechnology. 2015;**10**(6):517-521

[34] Tran V, Soklaski R, Liang Y, Yang L. Layer-controlled band gap and

anisotropic excitons in few-layer black phosphorus. Physical Review B. 2014; **89**(23):235319

[35] Li L, Kim J, Jin C, Ye GJ, Qiu DY, Felipe H, et al. Direct observation of the layer-dependent electronic structure in phosphorene. Nature Nanotechnology. 2017;**12**(1):21-25

[36] Kim J, Su Baik S, Ryu SH, Sohn Y, Park S, Park B-G, et al. Observation of tunable band gap and anisotropic Dirac semimetal state in black phosphorus. Science. 2015;**349**(6249):723-726

[37] Deng B, Vy T, Xie Y, Jiang H, Li C, Guo Q, et al. Efficient electrical control of thin-film black phosphorus bandgap. Nature Communications. 2017;**8**:14474

[38] Quereda J, San-Jose P, Parente V, Vaquero-Garzon L, Molina-Mendoza AJ, Agraït N, et al. Strong modulation of optical properties in black phosphorus through strain-engineered rippling. Nano Letters. 2016;**16**(5):2931-2937

[39] Zhang Z, Li L, Horng J, Wang NZ, Yang F, Yijun Y, et al. Strain-modulated bandgap and piezo-resistive effect in black phosphorus field-effect transistors. Nano Letters. 2017;**17**(10): 6097-6103

[40] Liu B, Köpf M, Abbas AN, Wang X, Guo Q, Jia Y, et al. Black arsenic–phosphorus: Layered anisotropic infrared semiconductors with highly tunable compositions and properties. Advanced Materials. 2015;**27**(30): 4423-4429

[41] Long M, Gao A, Wang P, Xia H, Ott C, Pan C, et al. Room temperature high-detectivity mid-infrared photodetectors based on black arsenic phosphorus. Science Advances. 2017; **3**(6):e1700589

[42] Yuan S, Shen C, Deng B, Chen X, Guo Q, Ma Y, et al. Air-stable room-temperature mid-infrared photodetectors based on hBN/black arsenic phosphorus/hbn heterostructures. Nano Letters. 2018; **18**(5):3172-3179

[43] Li L, Yang F, Ye GJ, Zhang Z, Zhu Z, Lou W, et al. Quantum hall effect in black phosphorus two-dimensional electron system. Nature Nanotechnology. 2016;**11**(7):593-597

[44] Li L, Ye GJ, Tran V, Fei R, Chen G, Wang H, et al. Quantum oscillations in a two-dimensional electron gas in black phosphorus thin films. Nature Nanotechnology. 2015;**10**(7):608-613

[45] Shoenberg D. Magnetic Oscillations in Metals. Cambridge Monographs on Physics. Cambridge University Press; 1984

[46] Svoboda P, Středa P, Nachtwei G, Jaeger A, Cukr M, Láznička M. Current-induced coupling of the edge and bulk channels in $GaAs/Al_xGa_{1-x}As$ heterostructures. Physical Review B. 1992;**45**(15):8763

[47] Maryenko D, Falson J, Bahramy MS, Dmitriev IA, Kozuka Y, Tsukazaki A, et al. Spin-selective electron quantum transport in nonmagnetic MgZnO/ZnO heterostructures. Physical Review Letters. 2015;**115**(19):197601

[48] Lin J, Han T, Piot BA, Wu Z, Xu S, Long G, et al. Determining interaction enhanced valley susceptibility in spin-valley-locked MoS_2. Nano Letters. 2019; **19**(3):1736-1742

[49] Shuigang X, Shen J, Long G, Zefei W, Bao Z-q, Liu C-C, et al. Odd-integer quantum hall states and giant spin susceptibility in p-type few-layer WSe_2. Physical Review Letters. 2017; **118**(6):067702

[50] Yang F, Zhang Z, Zhou Wang N, Jun Ye G, Lou W, Zhou X, et al. Quantum hall effect in electron-doped black phosphorus field-effect

transistors. Nano Letters. 2018;**18**(10): 6611-6616

[51] Yang J, Tran S, Wu J, Che S, Stepanov P, Taniguchi T, et al. Integer and fractional quantum hall effect in ultrahigh quality few-layer black phosphorus transistors. Nano Letters. 2018;**18**(1):229-234

[52] Long G, Xu S, Zhang T, Wu Z, Wong WK, Han T, et al. Charge density wave phase transition on the surface of electrostatically doped multilayer graphene. Applied Physics Letters. 2016; **109**(18):183107

[53] Ye JT, Zhang YJ, Akashi R, Bahramy MS, Arita R, Iwasa Y. Superconducting dome in a gate-tuned band insulator. Science. 2012;**338**(6111): 1193-1196

[54] Lu JM, Zheliuk O, Leermakers I, Yuan NF, Zeitler U, Law KT, et al. Evidence for two-dimensional Ising superconductivity in gated MoS_2. Science. 2015;**350**(6266):1353-1357

[55] Tran S, Yang J, Gillgren N, Espiritu T, Shi Y, Watanabe K, et al. Surface transport and quantum hall effect in ambipolar black phosphorus double quantum wells. Science Advances. 2017;**3**(6):e1603179

Chapter 3

Synthesis of Three-Dimensional Nanocarbon Hybrids by Chemical Vapor Deposition

Hua-Fei Li, Shuguang Deng and Gui-Ping Dai

Abstract

Carbon nanomaterials such as graphene, carbon nanotube (CNT), and carbon nanofiber (CNF) have received tremendous attentions in the past two decades due to their extraordinary mechanical strength and thermal and electrical properties. Recently, it indicates that three-dimensional (3D) nanocarbon hybrids overcome the weakness of individual low-dimensional nanocarbon materials and exhibit unique properties among carbon nanomaterials. Efforts have thus been made to acquire synergistic integration of one-dimensional (1D) and two-dimensional (2D) carbon nanomaterials. Meanwhile, chemical vapor deposition (CVD) is a wide-spread and effective method of fabricating three-dimensional nanocarbon hybrids compared with other synthetic methods. In this case, a number of 3D nanocarbon hybrids are synthesized by using different precursors at diverse temperature, and the nanocarbon hybrids are expected to be a promising choice for various application areas in the future.

Keywords: chemical vapor deposition, three-dimensional nanocarbon hybrids, graphene, carbon nanotube, carbon nanofiber

1. Introduction

As the typical 1D carbon materials, carbon nanotubes (CNTs) (**Figure 1a**) and carbon nanofibers (CNFs) have been widely investigated in the past two decades because of their merits, such as outstanding mechanical strength, large surface-to-volume ratio, and extraordinary electrical conductivity [1–4]. At the same time, graphene (**Figure 1b**), a recently discovered 2D sp^2 carbon, displays outstanding physical and chemical properties such as high specific surface areas (about 2620 $m^2\ g^{-1}$), great lightweight, and fast electron transport kinetics [5–7]. Nevertheless, their physical and chemical performances inevitably decrease compared to the theoretical prediction result from the existence of the van der Waals interaction, generating easy self-aggregation and stacking during the synthesis process [8]. Therefore, the 3D nanocarbon hybrids (such as CNT/graphene, CNF/graphene, CNT/CNF hybrids) are studied by a large number of the research groups, aiming at overcoming these shortcomings and a synergistic integration of their inherent properties in the new hybrid materials [9–11]. These nanocarbon hybrids have an interconnected network of carbon structure, resulting in a synergistic effect in enhanced conductivity in comparison with the individual components, and the special 3D structure significantly provides a variety of applications, such as field-effect transistors [12, 13], electron field emitters [14–16], sensors [17–20], fuel cell [21–23], batteries [24, 25], and supercapacitors [26–30].

To date, a number of techniques and methods have been utilized for the fabrication of nanocarbon hybrid, such as mixing process of surface-treated carbon materials (including solution processing [31, 32], vacuum filtration [33, 34], layer-by-layer self-assembly method [35, 36]), hydrothermal method [37], multi-step approaches using combinations of decorated carbon materials and CVD [10], and multi-step and one-step chemical vapor deposition [38–48]. Among all the nanocarbon hybrid fabrication approaches reported, CVD techniques are considered as the most versatile and promising way for nanocarbon composite production with reasonable structure and mechanical strength, which has attracted tremendous research attention during the recent decades. As a sophisticated synthesis method for both laboratory research and industry production, conventional CVD (shown in **Figure 2**) is applied in many areas, such as thin-film coating, crystal growth, and

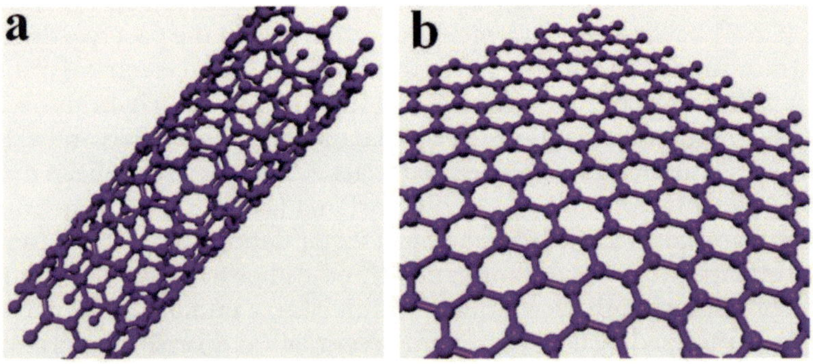

Figure 1.
Schematic diagrams of graphene and carbon nanotube.

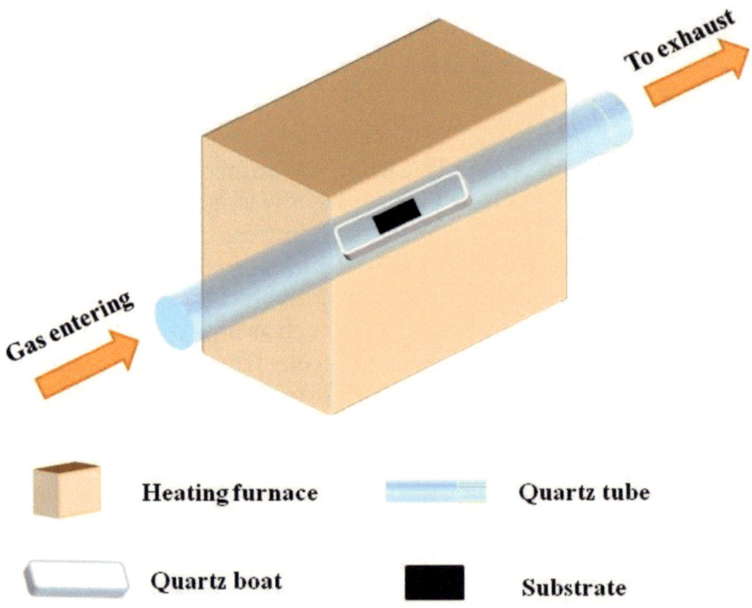

Figure 2.
Schematic diagrams of CVD technique.

powder production and also suitable for the synthesis of nanocarbon materials. The mechanism of conventional CVD generally includes two steps, initially thermal decomposition of gaseous precursor [10], organic solvents [47], or solid feedstock [9, 28] and thereafter reassembly of carbon atoms into sp^2 carbon nanostructures under the effective catalysis such as Cu [8], Fe [12], Co [28, 44], Ni [24], or their mixture [11, 49] at high temperature. Compared with other approaches, CVD technique significantly fabricated well-interconnected three-dimensional nanocarbon materials without needing sophisticated chemical routes with solvents and highly toxic agents during synthesis process. Zhu et al. [50] reported that the seamless, covalently bonded three-dimensional nanocarbon architecture was fabricated on the surface of Cu foil via simple two-step CVD methods. It is worth noting that although different aforementioned methods are employed for the production of nanocarbon hybrids, a facile and simple approach for controllable growth of three-dimensional carbonaceous nanomaterials is still a big challenge.

In this chapter, we present a summary of the researches about nanocarbon hybrid in recent years, with a focus on the popular fabrication techniques. Moreover, the merits and demerits and effect of experimental parameters of these CVD methods are presented in detail. Finally, we discuss the development trend, challenges, and performance applications of nanocarbon hybrids in the further.

2. Preparation techniques of three-dimensional nanocarbon hybrids

Up to now, varied approaches have been used for the fabrication of 3D nanocarbon hybrids, and the preparation technology generally could be categorized into four different approaches (shown in **Figure 3**): mixing process of surface-treated carbon materials, hydrothermal method, multi-step approaches using combinations of decorated carbon materials and CVD, and multi-step and one-step chemical vapor deposition. In addition, early researches on the construction of 3D hybrids focus on mixing process, which includes solution processing, vacuum filtration, and layer-by-layer self-assembly methods. Compared to other methods, hydrothermal route is an appropriate way to the mass preparation of graphene-carbon nanotube hybrids because of the easy operation and mild experimental environment. Moreover, the composites consisted of carbon nanotube and carbon nanofiber mainly produced by utilizing the multi-step approaches using combinations of decorated carbon materials and CVD method. Especially, multi-step and one-step

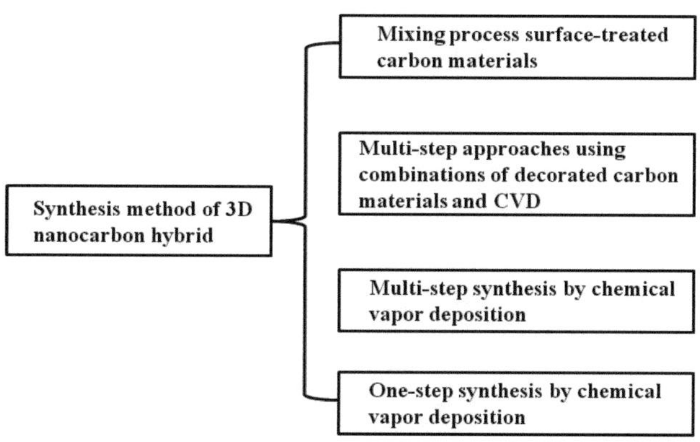

Figure 3.
Classification of 3D nanocarbon hybrids synthesis techniques.

chemical vapor deposition is considered as a simple and promising way to build 3D hybrids with hierarchical structure and stability.

2.1 Mixing process of surface-treated carbon materials

As the early hybridization approaches, solution processing, vacuum filtration, layer-by-layer self-assembly methods, and so on could be classified into the facile mixing process. Altogether 1D carbon nanomaterial incorporation of 2D nanomaterial with a facile mixing process exhibits a synergistic effect in enhanced properties. However, the nanocarbon hybrids are synthesized by utilizing various methods to mix modified carbon-based feedstocks, which generally need sophisticated chemical routes with solvents and highly toxic agents [31, 33–36]. Furthermore, this kind of techniques suffers from poor controllability, leading to the restriction of practical application due to the aggregation and stacking of carbon-based materials [51].

2.2 Hydrothermal method

With regard to hydrothermal method, firstly, the carbon feedstocks are dissolved and the mixed solution is transferred into a heating instrument. Secondly, the hydrothermal treatment is performed at low temperature, and the final product is obtained after centrifugation, washing, and freeze-drying process. Although this method has merits of mild conditions and scale-up synthesis, it is not suitable for the industrial production due to the time-consuming fault and defective products. Besides, the obtained 3D nanostructures are chiefly based on weak interconnection between individual nanocarbon components instead of owning powerful bonding, leading to robust 3D architecture [52].

2.3 Multi-step approaches using combinations of decorated carbon materials and CVD

Chemical vapor deposition is considered as the most promising approach of the preparation of graphene, CNTs, and CNFs on the substrate surface. Thus, it is always employed to facilitate the growth of CNTs on the decorated carbon materials, leading to the 3D hierarchical composite. For example, most of the reported 3D carbon nanotube/carbon nanofiber hierarchical composites are typically prepared by a multi-step route, which first needs electrospinning technique and post-carbonization for the preparation of CNFs, followed by decorating the CNFs with metal catalyst nanoparticles, and eventually the CNT growth is promoted by using toxic organic gases or solvent as carbon source during the CVD process [10, 22]. This kind of CVD-based methods has distinctive advantages: efficiency, convenience, and high yield. However, the stable and suitable decorated carbon materials that always need sophisticated pretreatment are vital to the construction of 3D carbon hybrids.

2.4 Multi-step synthesis by chemical vapor deposition

Multi-step chemical vapor depositions have been utilized in recent years to integrate individual 1D with 2D carbon nanomaterials to achieve controllable configurations of 3D nanostructures. Recently, Tang et al. successfully fabricated graphene-carbon nanotube composite on exfoliated vermiculite (EV) substrate by the multi-step CVD method (as shown in **Figure 4**). The whole CVD process could be divided into two steps: firstly, the aligned CNTs are synthesized at 650°C by using C_2H_4 as carbon source, and, secondly, the uniform graphene sheet directly grows on the surface of substrate at a higher temperature of 950°C by utilizing the hydrocarbon—CH_4, resulting in the in situ synthesis of graphene-carbon nanotube-graphene sandwiches [53]. In other methods of the successful fabrication of 3D hybrids, the obtained component materials are always entangled with each other, and the ordered 3D packing architecture is hardly available. Nevertheless, this multi-step way successfully integrates low-dimensional materials into 3D ordered, controllable, and well-connected structures [50]. Additionally, the morphology and nanostructure could be well controlled by adjusting the experimental parameters due to the separated CVD processes. It is a pity that the multi-step process always requires strict growth conditions and large consumption of power (high temperature) for the growth of nanocarbon materials.

2.5 One-step synthesis by chemical vapor deposition

Recently, tremendous efforts have been made to produce 3D nanocarbon hybrid via simultaneously in situ growing of 1D and 2D carbon nanomaterials on the surface of substrate during the CVD method. For example, Dong et al. (illustrated in **Figure 5**) reported that graphene/carbon nanotube hybrids were synthesized by a facile single-step CVD route employing ethanol (C_2H_5OH) as feedstock on the surface of Cu substrate decorated with Si nanoparticles, and the property and shape of hybrid could be varied by adjusting the fabrication environment (e.g., Si nanoparticles, temperature, and annealing time). The single-step route has the merits of better electrical conductivity and lesser defect density than the multi-step methods [43]. Additionally, although this one-step process effectively decreases the

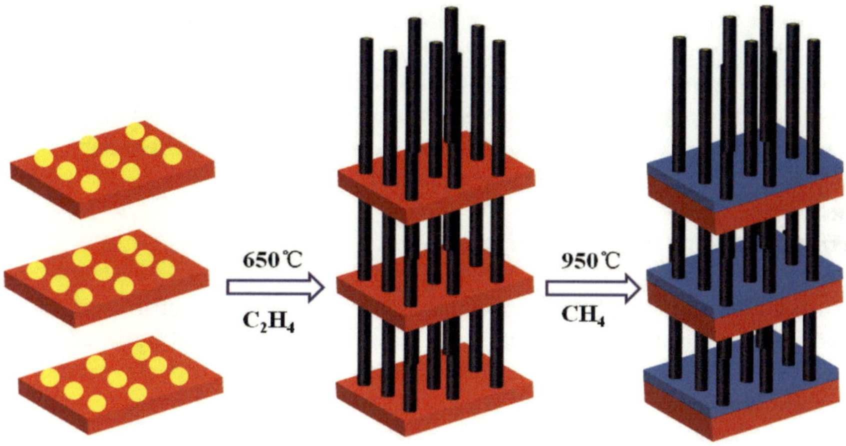

Figure 4.
Scheme for the two-step CVD synthesis of graphene/carbon nanotubes hybrids.

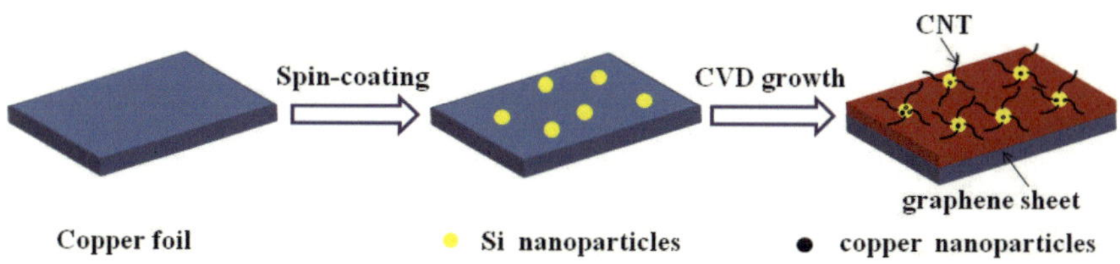

Figure 5.
Scheme for the one-step CVD synthesis of graphene/carbon nanotubes hybrids.

consumption of power, they still need high temperature, flammable gases, or toxic chemicals for the in-situ growth of 3D architecture.

3. Effect of experimental parameters of CVD methods

3.1 Effect of catalyst nanoparticles

It is known that substrate is the important part in the conventional CVD method, and the choice of substrate is essential to the morphology, nanostructure, and applications for carbonaceous nanomaterials. We generally use single transition metal substrate (Fe, Co, Ni, Cu, palladium (Pd) [41], ruthenium (Ru) [54]) as the catalyst for the preparation of graphene, and Fe, Co, Ni, and Cu are of great interest, because of the low cost and availability.

Remarkably, to build the uniform 3D architecture, single metal substrate is not enough for the CVD growth process. Hence, substrate embedded with metal nanoparticles serves as the bifunctional catalyst to facilitate the synthesis of different dimensional carbon materials, and the crucial issue for the in situ growth of 3D hybrids depends on the stability of catalyst nanoparticles during the deposition process. In CVD methods, the metal nanoparticles for the growth of hybrids could be obtained by a variety of ways, such as spin coating [47], electron evaporation [50], template etching [55, 56], and so on. Moreover, the covalent C–C bonding between different dimensional carbon materials, which is of paramount importance for 3D nanostructure, is probable to be achieved by such methods [57–59]. Nguyen et al. fabricated graphene/carbon nanotube composite by employing the Cu substrate-embedded Fe nanoparticles as the catalyst in the simple CVD approach [15]. In which Cu foil served as the template for the graphene sheet preparation. Additionally iron nanoparticles served as the catalyst for the CNT preparation. Besides, similar report indicated that the diameter, density, and quality of CNTs of composite could be defined by the size of the catalyst nanoparticles [45]. And various densities of catalyst nanoparticles had a different effect on the purity, thermal stability, and defects of 3D carbon hybrids [59].

3.2 Effect of growth temperature

Although low-dimensional carbon nanomaterials' nanostructure and diameter in 3D architecture is directly related to the size and nature of catalyst nanoparticles, it also could be indirectly determined by adjusting growth temperature in CVD technique. The different growth of CNF/CNT hybrid was fabricated due to the different carbon source decomposition and diffusion rate at various growth temperatures in the study of Park et al. [24]. Furthermore, the growth temperature is also crucial for the defects and properties of 3D carbon nanomaterials. Lin et al. [57] found that at different growing temperatures, the various architecture of sample could be produced by indirectly changing the number of layers of graphene and packing density of CNTs. And the ratio of the I_D/I_G (Raman spectroscopy analysis), defects, and surface area increased with the decrease of growth temperature, leading to the increased specific capacitance. As a result, it is crucial to seek the appropriate growth temperature for the growth of well-developed 3D composite.

3.3 Effect of carrier gas

In the CVD approach, hydrogen (H_2), argon (Ar), and nitrogen (N_2) are utilized for the growth of carbon materials in the high-temperature annealing process, and the influence of variety of gases in the conventional CVD process is different. As for Ar and N_2, they serve as the carrier gas to introduce the vapor into the CVD furnace under a suitable flow rate. As for H_2, it has multifunctional effects in the practical CVD environment. First, it is believed that H_2 removes surface impurities (such as S and P) and defects which can cause local variations of carbon solubility in the metal substrate in the high-temperature process [13, 60]. And it also enables the reduction process of metal oxides for producing enough catalyst nanoparticles at the high temperature [10, 22, 24]. Yan et al. [49] fabricated mesoscopic 3D composite comprised of graphene and CNTs under the effect of Ni-Co catalysts which was produced at 800°C in H_2 atmosphere. Unlike the conventional CVD synthesis of individual 1D or 2D carbon nanostructure, H_2 also plays an important role in building nanostructure of 3D hierarchical hybrids, especially for graphene/CNT composite. For example, there were two simultaneous reactions appearing during the construction of 3D graphene/CNT hybrids in the previous report [8]. On one time, the methane decomposed with the increasing temperature and thereafter facilitates the CNT growth out of islands of metal catalyst. Simultaneously, hydrogenation process appeared on the surface of graphene sheet (shown in **Figure 6**). In this process, graphene sheet was effectively etched under the atmosphere of H_2 and transformed into CH_4 at the point of connection with the catalyst nanoparticles (Ni nanoparticle + C graphene +$2H_2$ → Ni + CH_4) [61]. Furthermore, the morphology of the hybrids was adjusted via varying the H_2 flow rate to change the two contrary reactions in the CVD method. Consequently, the high density of CNTs grown on the surface of graphene sheet under the suitable flow rate of H_2, implying that the rate of H_2 etching optimized the 3D nanocarbon formation.

3.4 Effect of carbon source

So far, quite a few investigations have been dedicated to the fabrication of 3D carbonaceous hybrids by using various carbon sources, and studies have illustrated that the carbon sources can also be basically classified into the three categories: hydrocarbon compounds (CH_4 [45, 58], C_2H_2 [11, 50], C_2H_4 [10], C_3H_8 [43]), liquid

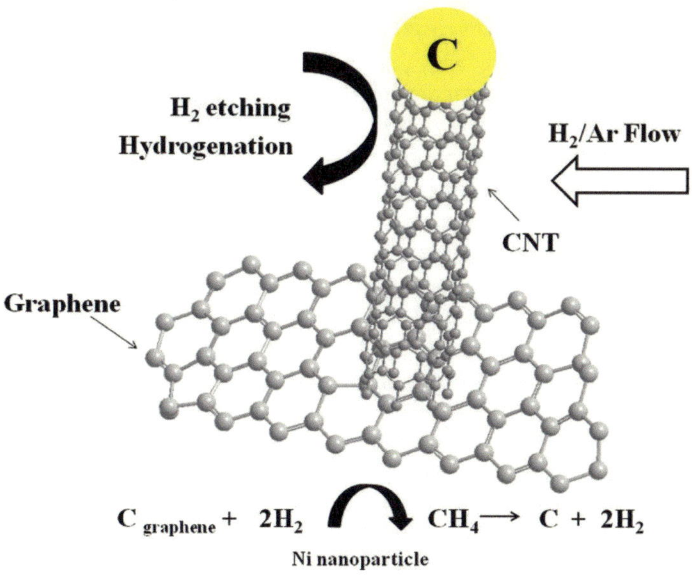

Figure 6.
Schematics illustrating direct CNT growth on planar graphene under H_2 etching.

carbon sources (ethanol [47], pyridine [22], toluene [41]), and solid feedstock (melamine [49], Prussian blue [9], camphor [62]) and so on. According to the relevant reports, diverse carbonaceous hybrids choose various carbon sources as feedstock for the basic supply of 3D architecture. With respect to CNT/CNF hybrids, hydrocarbon compounds are always considered as feedstocks of CNTs on the surface of obtained CNFs. With respect to CNT/graphene or CNF/graphene composites, hydrocarbon compounds, liquid carbon sources, and solid feedstock are all used as precursors for the growth of hierarchical architecture. Notably, for the synthesis of graphene, the present CVD technique requires high growth temperature, typically 1000°C [63–65]. Since it is more environment-friendly, convenient, and economical for industrial fabrication, a low-temperature route is greatly desirable. Liquid and solid carbon sources decompose at a lower temperature relative to major gaseous carbon sources. Therefore, liquid and solid feedstock could be a better choice for the growth of 3D CNT/graphene or CNF/graphene hybrids because of the quick carbon diffusivity through metal catalysts and covering on the surface at lower temperature. Moreover, during the dehydrogenation process of liquid or solid carbon sources, the overall dehydrogenation barrier and nucleation barrier are much lower than that of gaseous carbon source from the relevant report [66]. Recently, low temperature (800°C) one-step CVD synthesis of 3D hybrids composed by CNTs and graphene sheet are demonstrated by using melamine as the single solid carbon source [56]. Nevertheless, 3D hybrid growth at lower temperature still remains a challenge.

4. Development trend and application prospect of three-dimensional nanocarbon hybrids

Three-dimensional nanocarbon hybrids have been used for a variety of applications, for example, transparent and flexible electrodes and field-effect transistors [12–16, 47], sensors [17–20], fuel cell [67], batteries [9, 11, 44, 55], supercapacitors [10, 50, 51], and so on.

4.1 Three-dimensional nanocarbon hybrids in transparent and flexible electrodes and field-effect transistors

Because of the outstanding mechanical, electrical, and thermal properties, low dimensional nanocarbon materials have recently attracted enormous interest for potential application in transparent and flexible nanoelectronics [68–70]. Furthermore, 3D graphene-based hybrids which offset shortcomings of pure graphene received a large number of attentions in particular for two applications: transparent and flexible electrodes and field-effect transistors. Kim et al. [13] successful synthesized single-walled carbon nanotubes (SWCNT)/graphene hybrids on the Cu foil coated with CNTs. Notably, compared to pure CNT (58.78 ± 36.17 cm^2/V s) and graphene (341.7 ± 259.4 cm^2/V s), SWCNT-graphene hybrids possessed higher field-effect mobilities (μ) (394.46 ± 176.27 cm^2/V s) and better output characteristics (**Figure 7**), suggesting that the electrical conductivity of this hybrids dramatically increased compared to individual carbon material. As for transparent and flexible device applications, the hybrids showed the low sheet resistance (300 Ω/sq) with 96.4% optical transparency which is largely lower than the monolayer graphene (~1 kΩ/sq) grown by CVD method, indicating that composite is a promising material in developing high-performance transparent and flexible devices. Additionally, the hybrids possessed improved mechano-electrical property result from the CNT growth and obtained hybrid demonstrated that at an

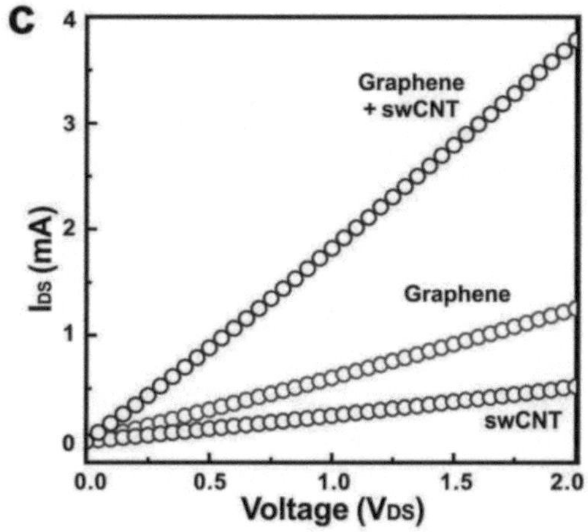

Figure 7.
Output characteristics ($I_{DS} - V_{DS}$) of graphene, SWCNT-graphene hybrid film, and SWCNT. Reproduced with permission from ref. [13].

applied field of 4.0 V/μm, the hybrid exhibited a current density of 1.33 mA cm^{-2} [15], implying superiority than that of pure CNT materials on indium tin oxide films (ITO) glass [71]. And it probably replaces the ITO films, the most common transparent and flexible electrodes, as an alternative material with properties including high on/off ratios and outstanding electrical conductivity for high-performance flexible device in the future.

4.2 Three-dimensional nanocarbon hybrids in fuel cell and batteries

In the practical application, the higher active and stable catalysts are crucial to the high electrochemical performance of fuel cell. Compared to the pure Pt-graphene cathode material, the Pt-3D nanocarbon composite cathode exhibits much smaller oxygen reduction reaction (ORR) charge transfer resistance and higher maximum power density in the direct methanol fuel cell [23] and proton exchange membrane fuel cells [21]. Moreover, due to the expensive cost and poor durability, as the spread anode and cathode electro-catalysts for ORR, Pt-based materials are hampered in the commercialization. Significantly, the CNT/CNF composite acts as the effective Pt-free ORR catalyst with a comparable activity, cheap price, and better thermal stability and durability, and the unique 3D network results in the enhanced electrochemical performance [22], implying 3D hybrid materials are becoming increasing competitive in the fuel cell applications.

Carbon-based materials (such as CNTs, CNFs, graphene), with their merits of reversible lithium-carbon reaction, low-intercalation potential with Li^+, high-coulumbic efficiencies, and low-capacity fade, are excellent choices as electrode materials of lithium-ion batteries [72, 73]. Nevertheless, the cycling performance and high-rate capability of individual material are not as satisfactory as expected, possibly owing to the large contact resistance of easy self-aggregation and stacking. Moreover, hybrids consisting of various low dimensional carbon materials, which favor different oriented diffusion of the lithium ion and the 3D nanocarbon architecture, are beneficial to the electrons' collection and transport around the cycling process, leading to high electrical conductivity and chemical stability. For example, the 3D nanocarbon hybrid anode exhibited significantly enhanced reversible capacity (300 mA h g^{-1}), outstanding cycling stability, and lower electrolyte resistance and contact resistances in contrast with the individual CNF material [11].

Additionally, by building 3D carbon network, at current densities of 0.36, 0.6, 1.2, 2.4, and 6 mA/cm^2, the rate performance of graphene/CNF hybrids reached about 420, 385, 329, 229, and 189 mA h g^{-1}, (as shown in **Figure 8**), which were superior to those of other pure nanocarbon performances [44]. Therefore, hybridization of the different low-dimensional carbon nanomaterials is an effective route to provide fast ion/electron transfer and higher Li storage capability, and the hierarchical 3D carbonaceous architecture is also promising for Li-ion battery applications in the future.

Because of the large energy density, capacity (1673 mA h g^{-1}), low cost, and environmental benignity of sulfur, lithium-sulfur (Li-S) batteries are investigated by a large number of research groups. However, the "shuttling effect" which always triggers an inevitable sulfur loss in practical Li-S battery applications, leading to an increase in internal resistance, low cycling capacity, and poor coulombic efficiency. To solve this problem, porous carbon materials, e.g., CNTs and graphene, also have been utilized to capture and encapsulate sulfur, blocking the high solubility of polysulfides during the Li-S battery applications [74, 75]. And compared to pure CNT which is always hindered by problems of easy self-aggregation, enormous interface resistance, and poor S-storage ability, the 3D hybrids composed of CNTs and graphene are more suitable for the cathode of high-rate performance for Li-S batteries. The hybrid structure exhibits unique advantages: (i) the well-connected junction between the CNTs and graphene sheets enable rapid electron transfer; (ii) robust nanostructure provides flexibility and mechanical robustness, which effectively buffers volume changes during the cycling process [9]. Zhao et al. reported that graphene/CNT composite cathode possessed remarkable performance: a reversible capacity (928 mA h g^{-1}) at 1 C capacity and at a high current rate of 5 C, the capacity as high as about 650 mA h g^{-1} could be obtained even after 100 cycles with a coulombic efficiency of about 92% in Li-S battery applications [76]. Furthermore, it is worth noting that electrochemical performance and catalytic activity have significantly improved nitrogen doping according to a relevant report, thus nitrogen-doped 3D hybrids also applied in the Li-S batteries. Tang et al. employed glucose and dicyandiamide as the carbon and nitrogen feedstocks to prepare the nitrogen-doped nanocarbon hybrid by a one-step chemical vapor deposition process technique, and the result (1314 mA h g^{-1} at 0.2 C, a capacity retention

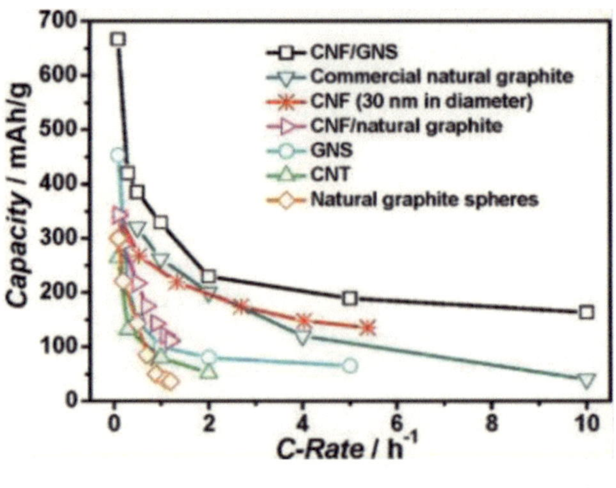

Figure 8.
Comparison of the rate capabilities of CNF/GNS, GNS, CNT, commercial natural graphite discharged at C/5, CNF (30 nm in diameter), CNF/natural graphite, and natural graphite spheres. Reproduced with permission from ref. [44].

of 97% after 200 cycles at a high rate of 2 C) exhibited the improved cyclic and rate performances [9]. These experimental results also indicate that the nitrogen-doped 3D nanostructure has the potential toward promising Li-S batteries.

4.3 Three-dimensional nanocarbon hybrids in supercapacitors

A variety of nanocarbon materials, e.g., CNTs, graphene, or mesoporous and activated carbon possess enormous specific surface areas yet are limited by low performance owing to aggregation and internal resistance, leading to decreased capacitance than theoretical prediction [77]. To overcome the aforementioned disadvantages, 3D composites are considered as attractive materials for supercapacitor application by inhibiting the agglomeration and improving the electrolyte electrode accessibility and the electrode conductivity. Relevant report demonstrated that the capacitance (653.7 μF cm^{-2}) at 10 mV s^{-1} of 3D CNT/graphene-based supercapacitor was superior to the graphene electrode (99.6 μF cm^{-2}) [8]. Zhou et al. fabricated supercapacitor based on polyaniline/carbon nanotube/carbon nanofiber (PANI/CNT/CNF) electrode [30]. Compared with pure PANI/CNF, the hybrids showed higher specific capacitance and energy density, superior rate capability, and lower ion diffusion/transport resistance (shown in **Figure 9**).

Meanwhile, due to the merits—high theoretical capacity, low cost, and natural abundance—diverse potential metal oxides, e.g., RuO_2 [78], MnO_2 [79], NiO [80],

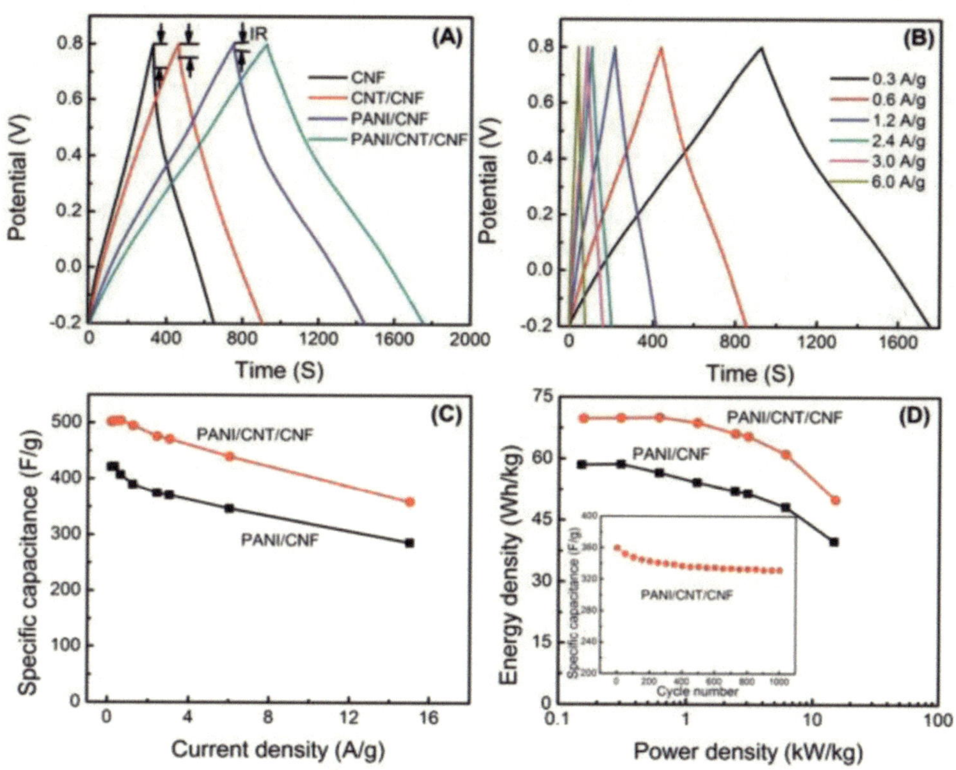

Figure 9.
(A) GCD curves of CNF, CNT/CNF, PANI/CNF, and PANI/CNT/CNF film electrodes at a current density of 0.3 a g⁻¹, respectively. (B) GCD curves of PANI/CNT/CNF film electrodes at different current densities. (C) Specific capacitance vs. current density for PANI/CNF and PANI/CNT/CNF film electrodes. (D) Charging/discharging cycling stability of PANI/CNT/CNF film electrodes at a current density of 15 a g⁻¹. Reproduced with permission from ref. [30].

and Co_3O_4 [81] are regarded as the potential materials for pseudocapacitors. Particularly, as one of the most promising pseudocapacitor materials, when the MnO_2 combined with 3D carbon hybrid, the drawbacks such as weak conductivity, low specific surface area, and brittleness of metal oxide electrodes are effectively alleviated, resulting in higher electrochemical performances [26, 27]. Wang et al. synthesized nanocarbon hierarchical composites (CNTs/CNFs) decorated with MnO_2 for flexible supercapacitors [10]. And the 3D nanocarbon hybrid/MnO_2 electrodes showed large better specific capacitance, cycling stability, maximum energy density, and rate capability than the CNF/MnO_2 electrodes. These enhanced electrochemical performances of hybridized-based electrodes indicate that the designed hierarchical structures of composites support a large special surface area for the reaction between electrolyte ions and metal oxides. Simultaneously, the special 3D nanostructures improve the electrode nanomaterials' electronic conductivity and facilitate transport channels for electrolyte ions. It is no doubt that 3D nanocarbon hybrids will have a crucial impact on the emerging materials of high-performance supercapacitor applications.

5. Conclusion and further prospects

To combine the merits of each building block, 3D nanocarbon structures (CNT/graphene, CNF/graphene, CNT/CNF hybrids) have been prepared by a variety of methods. The synthesis procedure, merits, and demerits of different approaches reported in the literatures are discussed in this chapter. Among them, chemical vapor deposition is regarded as the most promising fabrication method, and nicely hybrid architectures are achievable by this way. Nevertheless, there are various drawbacks and challenges in the practice synthesis. One of the great challenges in the CVD synthesis of the three-dimensional nanocarbon hybrids is convenience or simpleness when compared to preparation methods of individual nanocarbon materials. A simple scalable CVD method to fabricate controllable architecture of 3D nanocarbon hybrid is still crucial to industrial production. Furthermore, a variety of applications have been presented in this chapter. Compared to individual nanocarbon components, the superior performances of 3D nanocarbon hybrids signify their promising and wide application in the future, and 3D hybrid electrode materials are becoming more competitive in energy storage applications. It is worth mentioning that studies on the growth mechanism of 3D nanostructure which is necessary for the full understanding of CVD growth process is seldom reported from the relevant literatures. And some crucial problem still remained to be solved, particularly the interactions between various individual components and structure control in the future.

Author details

Hua-Fei Li[1], Shuguang Deng[2]* and Gui-Ping Dai[3]*

1 Institute for Advanced Study, Nanchang University, Nanchang, Jiangxi, China

2 Department of Chemical Engineering, School for Engineering of Matter, Transport and Energy, Arizona State University, Tempe, USA

3 Department of Chemical Engineering, School of Environmental and Chemical Engineering, Nanchang University, Nanchang, Jiangxi, China

*Address all correspondence to: shuguang.deng@asu.edu and nanodai@gmail.com

References

[1] Baughman RH, Zakhidov AA, de Heer WA. Carbon nanotubes-the route toward applications. Science. 2002;**297**(5582):787-792

[2] Ebbesen TW, Ajayan PM. Large-scale synthesis of carbon nanotubes. Nature. 1992;**358**(6383):220-222

[3] De Arco LG, Zhang Y, Kumar A, Zhou CW. Synthesis, transfer, and devices of single-and few-layer graphene by chemical vapor deposition. IEEE Transactions on Nanotechnology. 2009;**8**(2):135-138

[4] Hammel E, Tang X, Trampert M, Schmitt T, Mauthner K, Eder A, et al. Carbon nanofibers for composite applications. Carbon. 2004;**42**(5-6):1153-1158

[5] Huang X, Zeng ZY, Fan ZX, Liu JQ, Zhang H. Graphene-based electrodes. Advanced Materials. 2012;**24**(45):5979-6004

[6] Novoselov KS, Geim AK, Morozov SV, Jiang D, Katsnelson MI, Grigorieva IV, et al. Two-dimensional gas of massless Dirac fermions in graphene. Nature. 2005;**438**:197-200

[7] Xia XH, Chao DL, Fan ZX, Guan C, Cao XH, Zhang H, et al. A new type of porous graphite foams and their integrated composites with oxide/polymer core/shell nanowires for supercapacitors: Structural design, fabrication, and full supercapacitor demonstrations. Nano Letters. 2014;**14**(3):1651-1658

[8] Kim YS, Kumar K, Fisher FT, Yang EH. Out-of-plane growth of CNTs on graphene for supercapacitor applications. Nanotechnology. 2012;**23**:15301-15307

[9] Su DW, Cortie M, Wang GX. Fabrication of N-doped graphene-carbon nanotube hybrids from prussian blue for lithium-sulfur batteries. Advanced Energy Materials. 2017;**7**:1602014

[10] Wang T, Song DF, Zhao H, Chen JY, Zhao CH, Chen LL, et al. Facilitated transport channels in carbon nanotube/carbon nanofiber hierarchical composites decorated with manganese dioxide for flexible supercapacitors. Journal of Power Sources. 2015;**274**:709-717

[11] Kim H, Huang XK, Guo XR, Wen ZH, Cui SM, Chen JH. Novel hybrid carbon nanofiber/highly branched graphene nanosheet for anode materials in lithium-ion batteries. ACS Applied Materials & Interfaces. 2014;**6**:18590-18596

[12] Li B, Cao XH, Ong HG, Cheah JW, Zhou XZ, Yin ZY, et al. All-carbon electronic devices fabricated by directly grown single-walled carbon nanotubes on reduced graphene oxide electrodes. Advanced Materials. 2010;**22**(28):3058-3061

[13] Kim SH, Song W, Jung MW, Kang MA, Kim K, Chang SJ, et al. Carbon nanotube and graphene hybrid thin film for transparent electrodes and field effect transistors. Advanced Materials. 2014;**26**(25):4247-4252

[14] Deng JH, Cheng GA, Zheng RT, Yu B, Li GZ, Hou XG, et al. Catalyst-free, self-assembly, and controllable synthesis of graphene flake–carbon nanotube composites for high-performance field emission. Carbon. 2014;**67**:525-533

[15] Nguyen DD, Tai NH, Chen SY, Chueh YL. Controlled growth of carbon nanotube–graphene hybrid materials for flexible and transparent conductors and electron field emitters. Nanoscale. 2012;**4**(2):632-638

[16] Hong TK, Lee DW, Choi HJ, Shin HS, Kim BS. Transparent. Flexible conducting hybrid multilayer thin films of multi-walled carbon nanotubes with Graphene Nanosheets. ACS Nano. 2010;**4**(7):3861-3868

[17] Yan T, Wang Z, Wang YQ, Pan ZJ. Carbon/graphene composite nanofiber yarns for highly sensitive strain sensors. Materials & Design. 2018;**143**:214-223

[18] Unnikrishnan B, Mani V, Chen SM. Highly sensitive amperometric sensor for carbamazepine determination based on electrochemically reduced graphene oxide–single-walled carbon nanotube composite film. Sensors and Actuators B: Chemical. 2012;**173**:274-280

[19] Mani V, Devadas B, Chen SM. Direct electrochemistry of glucose oxidase at electrochemically reduced graphene oxide-multiwalled carbon nanotubes hybrid material modified electrode for glucose biosensor. Biosensors and Bioelectronics. 2013;**41**:309-315

[20] Chen X, Zhu J, Xi Q, Yang W. A high performance electrochemical sensor for acetaminophen based on single-walled carbon nanotube–graphene nanosheet hybrid films. Sensors and Actuators B: Chemical. 2010;**161**(1):648-654

[21] Yun YS, Kim D, Tak Y, Jin HJ. Porous graphene/carbon nanotube composite cathode for proton exchange membrane fuel cell. Synthetic Metals. 2010;**161**(21-22):2460-2465

[22] Guo QH, Zhao D, Liu SW, Chen SL, Hanif M, Hou HQ. Free-standing nitrogen-doped carbon nanotubes at electrospun carbon nanofibers composite as an efficient electrocatalyst for oxygen reduction. Electrochimica Acta. 2014;**138**:318-324

[23] Jha N, Jafri RI, Rajalakshmi N, Ramaprabhu S. Graphene-multi walled carbon nanotube hybrid electrocatalyst support material for direct methanol fuel cell. International Journal of Hydrogen Energy. 2011;**36**(12):7284-7290

[24] Park M, Jung YJ, Kim JY, Lee HI, Cho J. Synergistic effect of carbon nanofiber/nanotube composite catalyst on carbon felt electrode for high-performance all-vanadium redox flow battery. Nano Letters. 2013;**13**:4833-4839

[25] Dufficy MK, Khan SA, Fedkiw PS. Hierarchical graphene-containing carbon nanofibers for lithium-ion battery anodes. ACS Applied Materials & Interfaces. 2016;**8**(2):1327-1336

[26] Xiong CY, Li TH, Dang A, Zhao TK, Li H, Lv HQ. Two-step approach of fabrication of three-dimensional MnO_2 graphene-carbon nanotube hybrid as a binder-free supercapacitor electrode. Journal of Power Sources. 2016;**306**:602-610

[27] Lee DG, Kim BH. MnO_2 decorated on electrospun carbon nanofiber/graphene composites as supercapacitor electrode materials. Synthetic Metals. 2016;**219**:115-123

[28] Fan ZJ, Yan J, Zhi LJ, Zhang Q, Wei T, Feng J, et al. A three-dimensional carbon nanotube/graphene sandwich and its application as electrode in supercapacitors. Advanced Materials. 2010;**22**:3723-3728

[29] Zhao L, Qiu YJ, Yu J, Deng XY, Dai CL, Bai XD. Carbon nanofibers with radially grown graphene sheets derived from electrospinning for aqueous supercapacitors with high working voltage and energy density. Nanoscale. 2013;**5**:4902-4909

[30] Zhou ZP, Wu XF, Hou HQ. Electrospun carbon nanofibers surface-grown with carbon nanotubes and polyaniline for use as high-performance electrode materials of supercapacitors. RSC Advances. 2014;**4**(45):23622-23629

[31] Tung VC, Chen LM, Allen MJ, Wassei JK, Nelson K, Kaner RB, et al. Low-temperature solution processing of graphene-carbon nanotube hybrid materials for high-performance transparent conductors. Nano Letters. 2009;**9**(5):1949-1955

[32] King PJ, Khan U, Lotya M, De S, Coleman JN. Improvement of transparent conducting nanotube films by addition of small quantities of graphene. ACS Nano. 2010;**4**(7):4238-4246

[33] Tang Y, Gou J. Synergistic effect on electrical conductivity of few-layer graphene/multi-walled carbon nanotube paper. Materials Letters. 2010;**64**(22):2513-2516

[34] Khan U, O'Connor I, Gun'ko YK, Coleman JN. The preparation of hybrid films of carbon nanotubes and nano-graphite/graphene with excellent mechanical and electrical properties. Carbon. 2010;**48**(10):2825-2830

[35] Hong TK, Lee DW, Choi HJ, Shin HS, Kim BS. Transparent, flexible conducting hybrid multilayer thin films of multiwalled carbon nanotubes with graphene nanosheets. ACS Nano. 2010;**4**(7):3861-3868

[36] Kim YK, Min DH. Durable large-area thin films of graphene/carbon nanotube double layers as a transparent electrode. Langmuir. 2009;**25**(19):11302-11306

[37] Chen P, Xiao TY, Qian YH, Li SS, Yu SH. A nitrogen-doped graphene/carbon nanotube nanocomposite with synergistically enhanced electrochemical activity. Advanced Materials. 2013;**25**(23):3192-3196

[38] Gong QM, Li Z, Zhou XW, Wu JJ, Wang Y, Liang J. Synthesis and characterization of in situ grown carbon nanofiber/nanotube reinforced carbon/carbon composites. Carbon. 2005;**43**(11):2426-2429

[39] Thostenson ET, Li WZ, Wang DZ, Ren ZF, Chou TW. Carbon nanotube/carbon fiber hybrid multiscale composites. Journal of Applied Physics. 2002;**91**(9):6034-6037

[40] Gaikwad AV, Rout TK, Van der Plas D, Dennis RV, Banerjee S, Pacheco Benito S, et al. Carbon nanotube/carbon nanofiber growth from industrial by-product gases on low- and high-alloy steels. Carbon. 2012;**50**(12):4722-4731

[41] Lai C, Guo Q, Wu XF, Reneker DH, Hou H. Growth of carbon nanostructures on carbonized electrospun nanofibers with palladium nanoparticles. Nanotechnology. 2008;**19**(19):195303

[42] Feng JM, Dai YJ. Water-assisted growth of graphene on carbon nanotubes by the chemical vapor deposition method. Nanoscale. 2013;**5**(10):4422-4426

[43] Sahoo RK, Jeyapandiarajan P, Devi Chandrasekhar K, Daniel BSS, Venimadhav A, Sant SB, et al. Single-step synthesis of graphene-carbon nanofiber hybrid material and its synergistic magnetic behaviour. Journal of Alloys and Compounds. 2014;**615**:348-354

[44] Fan ZJ, Yan J, Wei T, Ning GQ, Zhi LJ, Liu JC, et al. Nanographene-constructed carbon nanofibers grown on graphene sheets by chemical vapor deposition: High-performance anode materials for lithium ion batteries. ACS Nano. 2011;**5**(4):2787-2794

[45] Van Chuc N, Thanh CT, Van Tu N, Phuong VTQ, Thang PV, Thanh Tam NT. A simple approach to the fabrication of graphene-carbon nanotube hybrid films on copper substrate by chemical vapor deposition.

Journal of Materials Science and Technology. 2015;**31**(5):479-483

[46] Lee DH, Kim JE, Han TH, Hwang JW, Jeon S, Choi SY, et al. Versatile carbon hybrid films composed of vertical carbon nanotubes grown on mechanically compliant graphene films. Advanced Materials. 2010;**22**(11):1247-1252

[47] Dong XC, Li B, Wei A, Cao XH, Chan-Park MB, Zhang H, et al. One-step growth of graphene–carbon nanotube hybrid materials by chemical vapor deposition. Carbon. 2011;**49**(9):2944-2949

[48] Su Q, Liang Y, Feng X, Müllen K. Towards free-standing graphene/carbon nanotube composite films via acetylene-assisted thermolysis of organocobalt functionalized graphene sheets. Chemical Communications. 2010;**46**(43):8279-8281

[49] Yan XL, Li HF, Wang C, Jiang BB, Hu HY, Xie N, et al. Melamine as a single source for fabrication of mesoscopic 3D composites of N-doped carbon nanotubes on graphene. RSC Advances. 2018;**8**(22):12157-12164

[50] Zhu Y, Li L, Zhang CG, Casillas G, Sun ZZ, Yan Z, et al. A seamless three-dimensional carbon nanotube graphene hybrid material. Nature Communications. 2012;**3**(1):1225

[51] Yu DS, Dai LM. Self-assembled graphene/carbon nanotube hybrid films for supercapacitors. The Journal of Physical Chemistry Letters. 2010;**1**(2):467-470

[52] Sakthinathan S, Kubendhiran S, Chen SM. Hydrothermal synthesis of three dimensional graphene-multiwalled carbon nanotube nanocomposite for enhanced electro catalytic oxidation of caffeic acid. Electroanalysis. 2016;**29**(4):1103-1112

[53] Tang C, Zhang Q, Zhao MQ, Tian GL, Wei F. Resilient aligned carbon nanotube/graphene sandwiches for robust mechanical energy storage. Nano Energy. 2014;**7**:161-169

[54] Mattevi C, Kim H, Chhowalla M. A review of chemical vapour deposition of graphene on copper. Journal of Materials Chemistry. 2011;**21**(10):3324-3334

[55] Ding YL, Kopold P, Hahn K, van Aken PA, Maier J, Yu Y. Facile solid-state growth of 3D well-interconnected nitrogen-rich carbon nanotube-graphene hybrid architectures for lithium–sulfur batteries. Advanced Functional Materials. 2016;**26**(7):1112-1119

[56] Li HF, Wu F, Wang C, Zhang PX, Hu HY, Xie N, et al. One-step chemical vapor deposition synthesis of 3D N-doped carbon nanotube/N-doped graphene hybrid material on nickel foam. Nanomaterials. 2018;**8**:700

[57] Lin CC, Lin YW. Synthesis of carbon nanotube/graphene composites by one-step chemical vapor deposition for electrodes of electrochemical capacitors. Journal of Nanomaterials. 2015;**2015**:1-8

[58] Zhang WL, Xie HH, Zhang RF, Jian MQ, Wang CY, Zheng QS, et al. Synthesis of three-dimensional carbon nanotube/graphene hybrid materials by a two-step chemical vapor deposition process. Carbon. 2015;**86**:358-362

[59] Zhao MQ, Peng HJ, Zhang Q, Huang JQ, Tian GL, Tang C, et al. Controllable bulk growth of few-layer graphene/single-walled carbon nanotube hybrids containing Fe@C nanoparticles in a fluidized bed reactor. Carbon. 2014;**67**:554-563

[60] Rümmeli MH, Rocha CG, Ortmann F, Ibrahim I, Sevincli H, Börrnert F, et al. Graphene: Piecing it together. Advanced Materials. 2011;**23**(39):4471-4490

[61] Ci LJ, Song L, Jariwala D, Elías AL, Gao W, Terrones M, et al. Graphene shape control by multistage cutting and transfer. Advanced Materials. 2009;**21**(44):4487-4491

[62] Shinde SM, Kalita G, Sharma S, Papon R, Yusop MZ, Tanemura M. Synthesis of a three dimensional structure of vertically aligned carbon nanotubes and graphene from a single solid carbon source. RSC Advances. **2014;4**(26):13355-13360

[63] Li X et al. Large-area synthesis of high-quality and uniform graphene films on copper foils. Science. 2009;**324**(5932):1312-1314

[64] Lee S, Lee K, Zhong ZH. Wafer scale homogeneous bilayer graphene films by chemical vapor deposition. Nano Letters. 2010;**10**(11):4702-4707

[65] Li XS, Cai WW, Colombo L, Ruoff RS. Evolution of graphene growth on Ni and Cu by carbon isotope labeling. Nano Letters. 2009;**9**(12):4268-4272

[66] Li ZC, Wu P, Wang CX, Fan XD, Zhang WH, Zhai XF, et al. Low-temperature growth of graphene by chemical vapor deposition using solid and liquid carbon sources. ACS Nano. 2011;**5**:3385-3390

[67] Li SS, Luo YH, Lv W, Yu WJ, Wu S, Hou PX, et al. Vertically aligned carbon nanotubes grown on Graphene paper as electrodes in lithium-ion batteries and dye-sensitized solar cells. Advanced Energy Materials. 2011;**1**(4):486-490

[68] Javey A, Guo J, Wang Q, Lundstrom M, Dai H. Ballistic carbon nanotube field-effect transistors. Nature. 2003;**424**(6949):654-657

[69] Kakade BA, Pillai VK, Late DJ, Chavan PG, Sheini FJ, More MA, et al. High current density, low threshold field emission from functionalized carbon nanotube bucky paper. Applied Physics Letters. 2010;**97**(7):073102

[70] Novoselov KS, Geim AK, Morozov SV, Jiang D, Zhang Y, Dubonos SV, et al. Electric field effect in atomically thin carbon films. Science. 2004;**306**(5696):666-669

[71] Yang H, Shang X, Li Z, Qu S, Gu Z, Xu Y, et al. Synthesis of large-area single-walled carbon nanotube films on glass substrate and their field electron emission properties. Materials Chemistry and Physics. 2010;**124**(1):78-82

[72] Obrovac MN, Christensen L. Structural changes in silicon anodes during lithium insertion/extraction. Electrochemical and Solid-State Letters. 2004;**7**(5):A93-A96

[73] Winter M, Besenhard JO, Spahr ME, Novák P. Insertion electrode materials for rechargeable lithium batteries. Advanced Materials. 1998;**10**(10):725-763

[74] Zhou G, Wang DW, Li F, Hou PX, Yin LC, Liu C, et al. A flexible nanostructured Sulphur–carbon nanotube cathode with high rate performance for Li-S batteries. Energy & Environmental Science. 2012;**5**(10):8901-8906

[75] Wang HL, Yang Y, Liang YY, Robinson JT, Li YG, Jackson A, et al. Graphene-wrapped sulfur particles as a rechargeable lithium–sulfur battery cathode material with high capacity and cycling stability. Nano Letters. 2011;**11**(7):2644-2647

[76] Zhao MQ, Liu XF, Zhang Q, Tian GL, Huang JQ, Zhu WC, et al. Graphene/single-walled carbon nanotube hybrids: One-step catalytic growth and applications for high-rate Li–S batteries. ACS Nano. 2012;**6**(12):10759-10769

[77] Zhang LL, Zhao XS. Carbon-based materials as supercapacitor electrodes. Chemical Society Reviews. 2009;**38**(9):2520-2531

[78] Wu ZS, Wang DW, Ren W, Zhao J, Zhou G, Li F, et al. Anchoring hydrous RuO_2 on graphene sheets for high-performance electrochemical capacitors. Advanced Functional Materials. 2010;**20**(20):3595-3602

[79] Lv P, Feng YY, Li Y, Feng W. Carbon fabric-aligned carbon nanotube/MnO_2/conducting polymers ternary composite electrodes with high utilization and mass loading of MnO_2 for super-capacitors. Journal of Power Sources. 2012;**220**:160-168

[80] Liu J, Jiang J, Bosman M, Fan HJ. Three-dimensional tubular arrays of MnO_2–NiO nanoflakes with high areal pseudocapacitance. Journal of Materials Chemistry. 2012;**22**(6):2419-2426

[81] Liu JP, Jiang J, Cheng CW, Li HX, Zhang JX, Gong H, et al. Co_3O_4 nanowire@MnO_2 ultrathin nanosheet core/shell arrays: A new class of high-performance Pseudocapacitive materials. Advanced Materials. 2011;**23**(18):2076-2081

Chapter 4

Hybrid Ion Exchangers

Amita Somya

Abstract

Hybrid ion exchangers are of recent origin in the field of ion exchange chemistry. They have shown excellent chemical, mechanical and thermal stability conversant to both organic and inorganic counterparts. Very recently, new classes of ion exchangers have been studied by combining surfactants and inorganic metal phosphates. This article highlights the salient features of metal phosphates as ion exchangers, various development stages with the modifications, with an emphasis on the recent developments in the field of analytical chemistry, particularly surfactant-based hybrid fibrous and non-fibrous metal phosphates as ion exchangers. Surfactants or surface-active agents when present in the matrix of inorganic metal phosphates not only enhance their ion-exchange capacity but, also the selective adsorption of metal ions. Therefore, these materials are of great importance in industrial and environmental applications.

Keywords: hybrid ion exchangers, stability, surfactants, metal phosphates, inorganic metal phosphates

1. Introduction

Analytical chemistry, broadly conceived, underlines and contributes to almost all branches of chemistry as an experimental science. It plays an important role in nearly all aspects of chemistry, such as, agricultural, clinical, environmental, forensic, manufacturing, metallurgical and pharmaceutical chemistry. The goal of a chemical analysis is to provide information about the composition of a sample of matter.

The discipline of analytical chemistry consists of qualitative and quantitative analyses. The former deals with the identification of elements, ions or compounds present in a sample, while the latter deals with the determination of how much of one or more constituents is present; whether the sample is solid, liquid, gas or a mixture. Analytical methods are ordinarily classified according to the property that is observed in the final measurement process. Some more important of these properties as well as the names of the methods based upon these properties are given in **Table 1**.

Prior to chemical analysis, separations are extremely important in analytical chemistry. The aim of an analytical separation is, usually, to eliminate or reduce interferences so that quantitative analytical information can be obtained about complex mixtures. There is a variety of separation methods that are in common use, including precipitation, distillation, solvent extraction, crystallization, dialysis, ion-exchange, chromatography, electrophoresis, field flow fractionation etc.

Of all the different types of separation methods, chromatography has the unique position of being applicable to all types of problems in all branches of science. This technique provides a very efficient method for the identification, separation, determination and purification of chemical compounds. It has undergone explosive growth in the last 30–40 years. The chromatographic technique was first invented by a Russian

S.N.	Technique	Property measured	Principal areas of application
1.	Gravimetry	Weight of pure analyte or compound of known stoichiometry	Quantitative for major or minor components
2.	Titrimetry	Volume of standard reagent solution reacting with the analyte	Quantitative for major or minor components
3.	Atomic and molecular spectrometry	Wavelength and intensity of electromagnetic radiation emitted or absorbed by the analyte	Qualitative, quantitative or structural for major down to trace level components
4.	Mass spectrometry	Mass of analyte or fragments of it	Qualitative or structural for major down to trace level components isotope ratio
5.	Chromatography and electrophoresis	Various physico-chemical properties of separated analytes	Qualitative and quantitative separations of mixtures at major to trace levels
6.	Thermal analysis	Chemical/physical changes in the analyte when heated or cooled	Characterization of single or mixed major/minor components
7.	Electro-chemical analysis	Electrical properties of the analyte in solution	Qualitative and quantitative for major to trace level components
8.	Radiochemical analysis	Characteristic ionizing nuclear radiation emitted by the analyte	Qualitative and quantitative at major to trace levels

Table 1.
Analytical techniques and principal applications.

S.N.	Technique	Stationary phase	Mobile phase	Format	Principal sorption mechanism
1.	Paper chromatography (PC)	Cellulose-water complex	Liquid	Planar	Partition (adsorption, ion-exchange, exclusion)
2.	Thin layer chromatography (TLC)	Silica, cellulose, ion-exchange resin, controlled porosity solid	Liquid	Planar	Adsorption (partition, ion-exchange, exclusion)
3.	Gas-liquid chromatography (GLC)	Liquid	Gas	Column	Partition
4.	Gas-solid chromatography (GSC)	Solid	Gas	Column	Adsorption
5.	High-performance liquid chromatography (HPLC)	Solid or bonded phase	Liquid	Column	Modified partition (adsorption)
6.	Size-exclusion chromatography (SEC)	Controlled porosity solid	Liquid	Column	Exclusion
7.	Ion-exchange chromatography (IEC)	Ion-exchange resin or bonded-phase	Liquid	Column	Ion-exchange
8.	Ion chromatography (IC)	Ion-exchange resin or bonded-phase	Liquid	Column	Ion-exchange
9.	Chiral chromatography (CC)	Solid chiral selector	Liquid	Column	Selective adsorption

Table 2.
A classification of the principal chromatographic techniques.

botanist Mikhil Tswett in 1906, at the University of Warsaw. He coined the term 'Chromatography' from the Greek words 'Chromatos' and 'graphy' which mean 'color' and 'to write' respectively. The International Union of Pure and Applied Chemists (IUPAC) has drafted a recommended definition of chromatography: 'Chromatography is a physical method of separation in which the components to be separated are distributed between two phases, one of which is a stationary phase, while the other is a mobile phase' [1]. Since its discovery, this technique has undergone tremendous modifications and nowadays various types of chromatographic techniques have been developed for separating almost any kind of given mixture, whether coloured or colourless into its constituents and to test the purity of these constituents. The applications of chromatography have extensively been used in the last 50 years, owing not only to the development of several new types of chromatographic techniques, but also due to the growing needs of the scientists for better methods of separating the complex mixtures or metal ions [2]. The different chromatographic methods are summarized in **Table 2**.

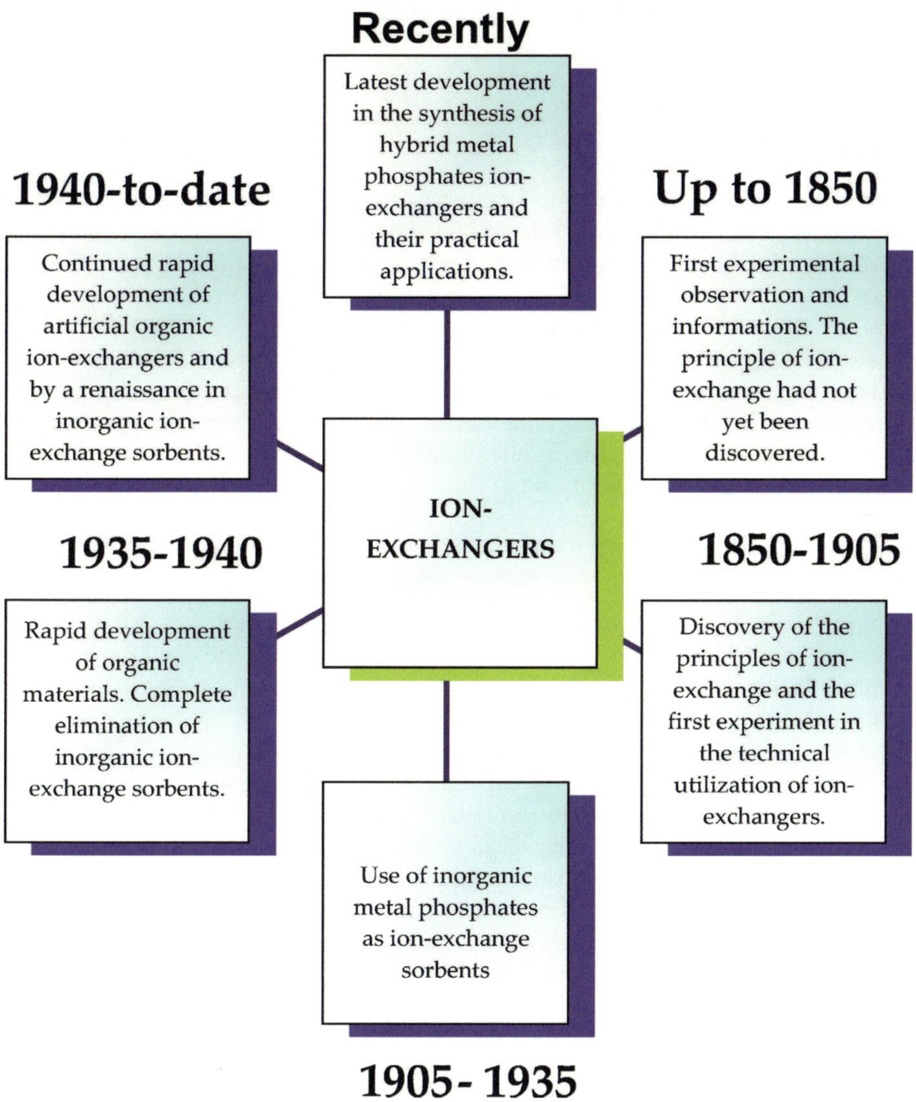

Figure 1.
Stages of development of ion exchangers and sorbents.

Out of these several chromatographic methods, ion-exchange has gained great attention by analysts in practice. The phenomenon of ion-exchange is not of recent origin. It has an interesting historical background. Various time spans may be identified for the development of ion-exchange technique. **Figure 1** summarizes the various stages of the development of ion exchangers and sorbents.

Initially, the ion exchangers were mostly used for water softening, but later on they were widely employed in various fields such as syntheses and some preparative works. The use of ion exchangers provided the new methods for analysts, which not only met the requirements of modern laboratories but also led to the solution of previously insolvable problems. Thus, the ion-exchange process has been established as an analytical tool in laboratories and industries. An interest in ion-exchange operations in industries is increasing day by day as their field of applications is expanding and today, it is an extremely valuable supplement to other procedures such as filtration, distillation and adsorption. All over the world, various plants are in operation, accomplishing tasks that range from the recovery of metals from industrial wastes to the separation of rare earths, and from catalysis of organic reactions to decontamination of water in cooling systems of nuclear reactors.

2. Ion exchangers

Ion exchangers are insoluble solid materials or immiscible liquids (in case of liquid ion exchangers) containing exchangeable ions. These ions can be exchanged for a stoichiometrically equivalent amount of other ions of the same sign on contacting with an electrolyte solution. Depending upon their ability of exchanging cations, anions or both, the ion exchangers may be categorized as 'cation', 'anion' or 'amphoteric' ion exchangers, respectively. A cation exchanger comprises a matrix with negative charge while an anion exchanger comprises a matrix with positive charge. The negative or positive charge on the matrix is compensated by the oppositely charged counter ions, which are mobile in nature. A typical ion-exchange reaction may be represented as follows:

$$R - A + B\ (aq) \rightleftharpoons R - B + A\ (aq) \tag{1}$$

where 'A' and 'B' are the replaceable counter ions, 'R' is the structural unit (matrix) of the ion exchanger and 'aq' stands for the aqueous phase. This process is reversible, that is, it can be reversed by suitably changing the concentration of the ions in solution.

The actual utility of an ion exchanger depends chiefly on its ion-exchange characteristics such as ion-exchange capacity, pH-titration, concentration, elution and distribution behaviour. The ion-exchange capacity depends on hydrated ionic radii and selectivity. The selectivity of any ion exchanger, in turn, is influenced by the nature of its functional group and degree of its cross linking. Ion exchangers, having groups that are capable of complex formation with some particular ions, will adsorb these ions more strongly. As the degree of cross linking increases, the exchanger becomes more selective towards ions of different sizes. The elution of H^+ ions from a column of ion exchanger depends on the concentration of the eluant while an optimum concentration of the eluant, necessary for maximum elution of H^+ ions, depends on the nature of ionogenic groups present in the exchanger, which depends upon the pKa values of the acids used in preparation. The efficiency of an ion exchanger depends on the following fundamental exchange reactions:

- Equivalence of exchange.
- Selectivity for one ion relative to another, including the cases in which the varying affinities of the ions are modified by the use of complexing and chelating agents.
- Donnan exclusion—the ability to exclude ions but not, in general, undissociated substances.
- Screening effect—the inability of very large ions or polymers to be adsorbed to an appreciable extent.
- Differences in migration rates of adsorbed substances down a column—primarily a reflection of differences in affinity.
- Ionic mobility restricted to the exchangeable ions and counter ions only.
- Miscellaneous properties—swelling, surface area and other mechanical properties.

On the basis of the nature of matrix, an ion exchanger may be 'organic' or 'inorganic' in nature.

2.1 Organic ion exchangers

Organic ion exchangers, commonly known as 'ion-exchange resins', are well known for their uniformity, chemical and mechanical stability and for the easy control over their ion-exchange property through synthetic methods. Organic resins have wide applications in analytical chemistry because of their high stability in the wide range of pH and reproducibility in the results, but their instability under the conditions of high temperature and strong radiation led to a major switch for the development of inorganic ion-exchange materials. The matrix of inorganic ion exchanger is more reactive than that of organic resins and hence, the selectivity for the metal ion depends both on adsorption characteristics of the matrix and the nature of the ionogenic groups attached to the matrix.

2.2 Inorganic ion exchangers

Inorganic ion exchangers are capable of being stable at elevated temperatures and in the presence of strong radiations and, hence, they have wide-ranging applications in nuclear researches such as radioisotope separations, nuclear waste treatments etc. They are used in the determination and detection of metals in pharmaceutical and biological products, analysis of alloys and rocks, as ion selective electrodes, as packing materials in ion-exchange chromatography and as catalysts. They also find applications in environmental analysis [3]. The widespread importance of inorganic ion exchangers in practical applications, and scientific interest in their nature and properties, has precipitated a wealth of published literature on the subject. Good starting points for further basic information are classic books like those of Clearfield [4], Amphlett [5] and Qureshi and Varshney [6]. These books have provided a complete picture and thorough insight into this field and widespread importance of inorganic ion exchangers. Important advances in this field have been reviewed by a number of workers/researchers at various stages of its development, such as Fuller [7], Qureshi et al. [8], Vesely and Pekarek [9], Clearfield [10, 11], Alberti et al. [12, 13], Alberti and Costantino [13], Marinsky [14], Varshney [15–20], Ivanov [21] and Terres-Rojas [22]. Dyer [23–25] has dealt with the theories involved zeolite molecular sieves, which have principles underlying the inorganic ion exchangers. Alberti (Itly) and Clearfield (USA) devoted most of their studies on the crystalline inorganic ion exchangers.

Inorganic ion exchangers are generally the oxides, hydroxides and insoluble acid salts of polyvalent metals, heteropolyacid salts and insoluble metal ferrocyanides. These materials are generally produced by combining the oxides of elements of III, IV, V and VI groups of the periodic table. A large number of such materials have been synthesized by mixing phosphoric, arsenic, molybdic, antimonic and vanadic acids with titanium, zirconium, tin, thorium, cerium, iron, antimony, chromium, niobium, tantalum, bismuth, nickel, cobalt, etc. However, the majority of work has been carried out on zirconium, titanium, tin, niobium and tantalum. Out of the above, metal phosphates have been found to have good chemical stability, reproducibility in ion-exchange behaviour and selectivity for certain metal ions.

2.3 Hybrid ion exchangers

Since organic ion exchangers were found to be unstable at elevated temperatures and under strong radiations, inorganic ion exchangers were taken as alternatives for such cases. However, the main drawback of inorganic ion-exchange materials has been that they are not very much reproducible in ion-exchange behaviour. Further, they are found not to be chemically and mechanically very stable perhaps due to their inorganic nature. Thus, to overcome these shortcomings, an interest was developed to obtain some organic-based inorganic ion exchangers. These exchangers were termed as 'hybrid ion exchangers' as they consist of both the organic and inorganic counterparts and have the properties not seen in purely organic or purely inorganic materials. This new class of ion exchangers has been prepared in these laboratories by incorporating a polymeric or monomeric organic species into the inorganic ion-exchange matrix [26–30]. The hybrid ion exchangers have shown an improvement in a number of ways. One of them is its granulometric properties that make it more suitable for the application in column operations. The binding with an organic species also introduces better mechanical properties in the end product, that is, hybrid ion-exchange materials. Hybrid ion exchangers can be prepared as three-dimensional porous materials in which layers are cross linked or as layered compounds containing sulphonic acid, carboxylic acid or amino groups.

The reactivity of both organic and inorganic precursors is usually quite different and phase separation tends to occur. The properties of hybrid materials do not depend only on organic and inorganic components but also on the interface between both phases. The general tendency is therefore, to increase interfacial interactions by creating an intimate mixing, or interpenetration between organic and inorganic networks. Moreover, the formation of chemical bonds between organic and inorganic species would prevent phase separation, allowing the synthesis of molecular composites or organic-inorganic copolymers. Hybrid materials can, thus, be divided into two classes [31].

- Class I corresponds to hybrid systems in which weak interactions such as van der Waals forces or hydrogen bonds or electrostatic interactions are created between organic and inorganic phases. This class involves mainly small organic species embedded within an oxide matrix.

- Class II corresponds to hybrid compounds where both organic and inorganic components are bonded through strong covalent chemical bonds.

2.4 Hybrid metal phosphates

Hybrid organo-inorganic phosphates open up a land of opportunities in materials science and ion-exchange chemistry. These nanocomposites bridge high-temperature materials such as glasses and ceramics with very fragile species such as organic compounds or biomolecules. In last 10–15 years, some hybrid ion-exchange materials have been synthesized in the laboratories, such as acrylamide

S.N.	Name of the materials	Ion-exchange capacity for Na^+ (meq/dry g)	Selectivity	X-ray nature	References
1.	Polyacrylonitrile thorium(IV) phosphate	3.90	Pb(II)	Microcrystalline	[36]
2.	Polystyrene cerium(IV) phosphate	2.95	Hg(II)	Microcrystalline	[37]
3.	Polystyrene thorium(IV) phosphate	4.52	Cd(II)	Crystalline	[38]
4.	Acrylonitrile cerium (IV) phosphate	2.86	Hg(II)	Poorly crystalline	[39]
5.	Acrylamide cerium(IV) phosphate	2.60	Hg(II)	Crystalline	[40]
6.	Acrylamide thorium(IV) phosphate	2.00	Pb(II)	Poorly crystalline	[41]
7.	Pectin cerium(IV) phosphate	1.78	Hg(II)	Amorphous	[42]
8.	Pectin thorium(IV) phosphate	2.15	Pb(II)	Amorphous	[42]
9.	Cellulose acetate thorium(IV) phosphate	1.70	Pb(II)	Amorphous	[43]
10.	Pyridine cerium(IV) phosphate	2.00	Hg(II)	Amorphous	[44]
11.	Pyridine thorium(IV) phosphate	2.10	Pb(II)	Amorphous	[45]
12.	n-Butyl acetate cerium(IV) phosphate	2.25	Hg(II)	Amorphous	[46]

Table 3.
Different types of hybrid metal phosphates and their important properties.

and pyridine-based zirconium and tin phosphates [32–34], acrylonitrile-based zirconium phosphate [35]. These materials have shown promising ion-exchange characteristics and have been utilized in the separation of metal ions due to their selectivity towards different metals ions. Metal phosphates, such as tin(IV) phosphate, cerium(IV) phosphate, zirconium(IV) phosphate etc. were found very good ion exchangers and intercalating agents too. The whole idea to convert them to hybrid ion exchangers has been to enhance the interlayer distances by introducing organic species in the metrics of these metal phosphates, resulting in improved ion-exchange properties. These metal-organic phosphates, or hybrid ion exchangers, correspond to metal-organic frameworks where features of organic and inorganic counterparts are revealed in terms of ion-exchange capacity, thermal, chemical and mechanical stability in addition to metal ion selectivity. However, structures of these metal-organic phosphates could not be described as these materials have been found amorphous or poorly crystalline. **Table 3** summarizes their ion-exchange capacities and selectivity towards metal ions.

3. Surfactants

Surfactants constitute the most important group of detergent components. 'Surfactant' is an abbreviation for surface-active agents (shown in **Figure 2**), which literally means 'active at the surface'. The surface can be between solid and liquid, between air and liquid and between two different immiscible liquids. The unique property of surfactants is 'adsorption', which occurs at liquid/solid, liquid/liquid and at air/liquid interfaces. At air-water interfaces and in water, or similarly strongly hydrogen-bonded solvents, they self-associate at concentrations above the critical micelle concentration (CMC) to form association colloids, known as 'micelles' [47].

Thus, 'surfactants, surface-active agents [48] or, detergents are amphiphilic, organic or organo-metallic compounds which form association colloids or micelles in solution'. Amphiphilic substances or amphiphiles are comprised of a hydrophobic portion, usually a long alkyl chain, attached to hydrophilic or water solubility-enhancing functional groups. Actually, surfactant molecule consists of two parts: a water-hating (hydrophobic) part and a water-loving (hydrophilic) part. The following figure shows the basic structure of a surfactant molecule.

There are three basic concepts that need to be well understood in order to explain the majority of observed phenomena; these are solubility, adsorption of a surfactant at a surface and the formation of micelles in solution. These three phenomena differentiate a surfactant from other chemical entities. It is the abnormal solubility characteristics of surfactants which offers the adsorption on surfaces/interfaces and formation of micelles. Surfactants reduce surface tensions when dissolved in water or water solutions, reduce interfacial tensions between two liquids, or between a liquid and a solid [49]. When a surfactant molecule is introduced into water, the water-hating

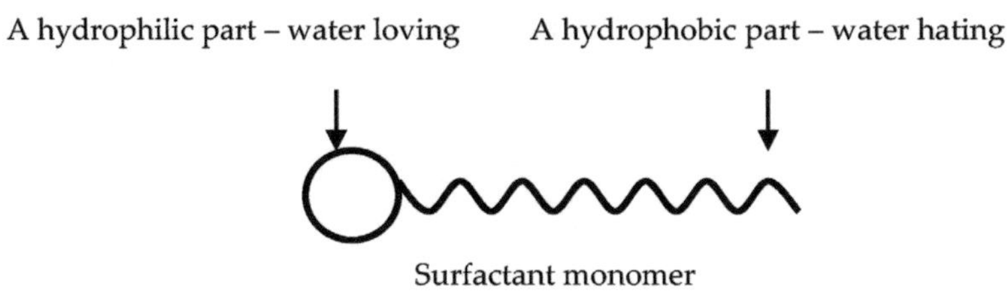

Figure 2.
Structure of Surfactant monomer.

part tends to escape by attaching itself to any available surface other than water. At the same time, the water-loving part tries to remain in water. As a result, surfactants get strongly 'adsorbed' to many surfaces, such as fabric, soil, glass and where the water and air meet (i.e., water/air interface). This tendency of surfactants is useful:

- In the removal of the soils from surfaces.
- In holding soil particles in suspension form and preventing them from redepositing onto the surface.
- In reducing surface tension of water and allowing the water to spread out.

The unusual properties of aqueous surfactant solutions can be ascribed to the presence of a hydrophilic head group and a hydrophobic chain (or tail) in the molecule. The polar or ionic head group usually interacts strongly with an aqueous environment, in which case it is solvated via dipole-dipole or ion-dipole interactions. Depending on the chemical structure of the hydrophilic moiety bound to the hydrophobic portion, the surfactants may be categorized into following types:

- Anionic surfactants [50]
- Cationic surfactants [51]
- Nonionic surfactants [52]
- Amphoteric surfactants [53]
- Gemini surfactants [54]

3.1 Micelle formation and critical micelle concentration (CMC)

The formation of micelles in aqueous solution is generally viewed as a compromise between the tendency for alkyl chains to avoid energetically unfavourable contacts with water, and the desire for the polar part to maintain contact with aqueous environment. In dilute aqueous solution, at concentration generally less than 10^{-4} M, the behaviour of ionic surfactants parallels that of strong electrolytes while the behaviour of nonionic surfactants often resembles that of the simple organic molecules. At higher surfactant concentrations, however, a pronounced deviation from 'ideal' behaviour in dilute solution occurs—this deviation generally being considerably larger than that exhibited by simple strong electrolytes. Thus, the adsorption of a surfactant from solution onto a surface depends upon the concentration [53]. Each surfactant has a characteristic CMC value. The most obvious evidence of micellar growth is probably the dramatic increase in viscosity with increasing concentration, which is observed in several surfactant solutions. Micellar growth is favoured by decreasing the temperature, adding electrolyte and lengthening the surfactant chain length and is, furthermore, very sensitive to the nature of the counter ion.

The physico-chemical properties of surfactants vary significantly below and above the CMC value [55]. Below the CMC value, the physico-chemical properties of ionic surfactants resemble those of a strong electrolyte. Above the CMC value, these properties change dramatically, indicating a highly co-operative association process. The general way of obtaining the CMC value of a surfactant micelle is to plot an appropriate physico-chemical property versus the surfactant concentration and observe the break in the plot [56–58].

Depending upon the chemical structure of the surfactant, its micelle can be cationic, anionic, zwitterionic or nonionic. The electrostatic character of the micelles depends in some cases on the pH of the aqueous solution due to protonation equilibria. Zwitterionic surfactants, of course, also can become either cationic or anionic, and several types of nonionic surfactants can also form anionic or cationic micelles in the appropriate pH range. Micelles are not static species but rather exist in a dynamic equilibrium. The micelle may be represented as a globular, cylindrical or ellipsoidal cluster [59] of individual surfactant molecules in equilibrium with its monomer. The reverse orientation of the hydrophilic and hydrophobic part of the surfactant in a hydrocarbon medium leads to the formation of reversed micelles [60].

3.2 Surfactant adsorption and surface properties

The effectiveness of surfactant adsorption is mainly determined by surfactant concentration, surfactant functional group, alkyl hydrocarbon chain length, environment etc. When the surfactant concentration is well below the CMC value, individual surfactant molecules tend to adsorb on exposed interfaces to reduce surface tension. As the concentration of surfactant approaches the CMC value, surfactant molecules form dimers, and multiple molecules aggregate, micelles. Once the CMC is reached, any additional surfactant molecules added to the system will be incorporated into new or existing aggregates. Thus, further increase of surfactant concentration above the CMC value results in bilayer or multilayer formation at interface. The adsorption tendency of the surfactants at the surfaces imparts the properties of foaming, wetting, emulsification, dispersing of solids and detergency. The adsorption increases the concentration of surfactant at the surfaces. Surfactant adsorption is a consideration in any application where surfactants come in contact with a surface or interface. It is from solutions that surfactants then preferentially adsorb to interfaces and, because of their amphiphilic nature, preferentially segregate at interfaces. There are a number of areas of applications where surfactant adsorption is important including ore floatation, improved oil recovery, soil remediation, detergency, surfactant-based separation processes and wetting. Surfactant adsorption may occur due to electrostatic interactions, van der Waals interaction, hydrogen-bonding and/or solvation and desolvation of adsorbate and adsorbent species [61].

The hydrophobic forces that drive surfactants to segregate at air-water interfaces are essentially the same that drive surfactant adsorption onto solid surfaces. However, they can be differing in the chemical forces associated with the solid surfaces. Ionic surfactants tend to adsorb onto oppositely charged solid surfaces due to electrostatic forces while adsorption of ionic surfactants on a like charged substrate, being less understood, can occur via hydrogen-bonding or attractive dispersion forces [62], as in the case for nonionic surfactants.

The number of industrial applications of surfactants is huge, and represents the subject of several book series. However, in analytical chemistry, surfactants have been recognized as being very useful in improving analytical technology, for example, in chromatography [63] and luminescence spectroscopy [64]. The use of surfactants in chromatography, particularly in ion-exchange, is of our interest. It is well known that surfactants are composed of two parts—hydrophobic and hydrophilic, which are oppositely charged, and the surfactants also act as ion exchangers. When their solutions are kept in contact with solid material, they are adsorbed on the solid surface with their hydrophilic part remaining in the solutions due to their surface-active property and they make that surface 'active'. It is clear from the above discussion that they play an important role in the adsorption behaviour of the molecules.

3.3 Surfactant-based hybrid ion exchangers

The widespread utility of surfactants in practical applications and scientific interest regarding their nature and properties have precipitated a wealthy literature [65, 66] on the subject. One of the predominant reasons for the ubiquitous applications of surfactants has been their remarkable ability to influence the properties of the surfaces and interfaces. Surfactants are widely used in various industrial applications [67–70] such as petroleum, pharmaceuticals, agro-chemicals, processing of foods, paints, coatings, adhesives, lubricants, in photographic films, personal care and laundry products.

Varshney et al. [71, 72] used surfactants as media in the adsorption studies of some alkaline earths and heavy metal ions on inorganic and hybrid ion exchangers and observed that the presence of surfactants in aqueous media increases the adsorption of metal ions on the surface on ion-exchange materials. Hence, exceptional high adsorption of the said metal ions has opened the doors in the field of material science. It was thought worthwhile to incorporate the surfactants in the matrix of inorganic ion exchangers to see how they could change the characteristics of the ion exchangers. Very recently, some hybrid fibrous and non-fibrous metal phosphates have been synthesized by combining surfactants and inorganic ion exchangers (metal phosphates) [73–81] by the researchers. Surfactants based ion exchangers also correspond to metal organic frameworks as surfactants being an organic counterpart introduced in the inorganic metal phosphates by the bonding in between the layers of metal phosphates. At this level too, structure could not been explained, reason being the amorphous and poorly crystalline nature of these materials.

Somya et al. [73–75] have probably first used surfactants in the synthesis some novel hybrid fibrous and non-fibrous metal phosphates by introducing surfactants (anionic, cationic and nonionic) in the matrix of inorganic metal phosphates. They have explored the ion-exchange studies such as ion-exchange, pH-titration, concentration, elution and thermal behaviour in addition to adsorption studies for some alkaline earths and heavy metal ions. Those materials were found to be selective for certain metal ions and, on that basis, some binary separations have been performed in the laboratory providing their potential role in environmental and analytical chemistry. The introduction of surfactants in the matrix of inorganic ion exchangers has been characterized by some physico-chemical studies like, IR, X-ray diffraction, elemental, SEM and TGA/DTA/DTG studies.

Later on, Iqbal [78–81] have synthesized the same class of hybrid metal phosphates by combining sodium dodecyl benzene sulphonate and sodium bis (2-ethylhexyl) sulphosuccinate in the matrix of cerium (IV) and tin (IV) phosphates. They have explored some ion-exchange studies in addition to physico-chemical characterization like IR, XRD, SEM, TGA/DTA/DTG, elemental studies and differential pulse polarography. These materials have shown selective adsorption for certain metal ions. Hence, binary separations have been done by using columns of the synthesized materials. Most of the surfactant based metal phosphates were found amorphous or poorly crystalline.

4. Conclusion

As per the studies done, so far, it is clearly indicated that surfactants have played a key role in synthesis of new class of hybrid metal phosphates as ion-exchange materials. They have enhanced not only the ion-exchange capacities of the inorganic metal phosphates when present in their matrix but also the adsorption of metal ions. These metal-organic hybrid materials have shown selectivity towards certain metals. Hence, these materials open a door with ample opportunities for the researchers in the field of analytical and environmental science where they can be used in water pollution control.

Acknowledgements

The author (A. Somya) is highly thankful to the Department of Applied Chemistry, Aligarh Muslim University, Aligarh for providing the research facilities. The author is also thankful to the Council of Scientific and Industrial Research, New Delhi for providing funds under Senior Research Fellowship scheme [09/112(0408)2K8-EMR-I]. The author wants to express special thanks to the Chancellor, Pro-Chancellor, Vice-Chancellor and Dean, School of Engineering, Presidency University, Bengaluru for giving enthusiastic and constant support.

Author details

Amita Somya
Department of Chemistry, School of Engineering, Presidency University, Bengaluru, Karnataka, India

*Address all correspondence to: somya.amita@gmail.com

References

[1] Ettre LS. Nomenclature for Chromatography. Pure and Applied Chemistry. 1993;**65**(4):819

[2] Qureshi M, Zafar S, Ahmad TS, Rahman N. Chemia Analityczna. 1992;**37**(1):21

[3] Varshney KG. New development in ion-exchange. In: Abe M, Kataoha T, Kodansha ST, editors. Proceeding Int. Conf. Ion-exchange (ICEE-91). Tokyo: Elsevier; 1991. p. 413

[4] Clearfield A. Inorganic Ion-exchange Materials. Boca Raton, Florida: CRC Press, Inc.; 1982

[5] Amphlett CB. Inorganic Ion-Exchangers. Amsterdam: Elsevier; 1964

[6] Qureshi M, Varshney KG. Inorganic Ion-exchangers in Chemical Analysis. Boca Raton, Florida: CRC Press, Inc.; 1991

[7] Fuller MJ. Inorganic ion-exchange chromatography on oxides and hydrous oxides. Chromatographic Reviews. 1971;**14**:45

[8] Qureshi M, Qureshi SZ, Gupta JP, Rathore HS. Recent progress in ion-exchange studies on insoluble salts of polybasic metals. Separation Science. 1972;7:615

[9] Vesely V, Pekarek V. Synthetic inorganic ion exchangers-1: Hydrous oxides and acidic salts of multivalent metals. Talanta. 1972;**19**:1245

[10] Clearfield A. Role of ion exchange in solid-state chemistry. Chemical Reviews. 1988;**88**:125

[11] Clearfield A. Inorganic ion exchangers: A technology ripe for development. I and EC Research. 1995;**34**:2865

[12] Alberti G, Casciola M, Costantino U. Inorganic ion-exchange membranes made of acid salts of tetravalent metals. A short review. Journal of Membrane Science. 1983;**16**:137

[13] Alberti G, Costantino U. Recent progress in the field of synthetic inorganic exchangers having a layered or fibrous structure. Journal of Chromatography. 1974;**102**:5

[14] Marinsky JA. Reactive and Functional Polymers. 1995;**27**(2):107

[15] Varshney KG, Gupta U. Tin(IV) antimonate as a lead-selective cation exchanger: Synthesis, characterization, and analytical applications. Bulletin of the Chemical Society of Japan. 1990;**63**:1515

[16] VarshneyKG,MohammadA.Chemical and Environmental Research. 1992;**1**(4):353

[17] Varshney KG. In: Sinha AK, Srivastava KN, Sinha BK, Pandey SK, Agarwal NK, editors. Environment Management in Developing Countries, Water and its Managements. Vol II; 1993. p. 64

[18] Varshney KG. Chemical and Environmental Research. 1994;**3&4**:301

[19] Varshney KG. In: Srivastava MM, Srivastava S, editors. Recent Trends in Chemistry. New Delhi: Discovery Publishing House; 2003. p. 467

[20] Varshney KG. Solid State Phenomena. 2003;**90-91**:445

[21] Ivanov VA, Timofeevskaya VD, Gorshkov VI, Drozdeva NV. Journal of Radioanalytical and Nuclear Chemistry. 1996;**208**(1):23

[22] Terres-Rojas E, Dominguez JM, Sales MA, Lopez E. Chemical Industries. 1997;**69**:567

[23] Dyer A. An Introduction to Zeolite Molecular Sieves. Avon: Bath Press Ltd.; 1988

[24] Dyer A, Townsend RP. The mobility of cations in synthetic zeolites with the faujasite framework — V: The self-diffusion of zinc into X and Y zeolites. Journal of Inorganic and Nuclear Chemistry. 1973;**35**:3001

[25] Dyer A, Hudson MJ, Williams PA. Progress in ion-exchange: Advances and applications. In: Proceedings of the Ion-Ex '95 Conference. Vol. 196. Special Publication by Royal Society of Chemistry; 1997. p. 498

[26] Chetverina RB, Boichinova ES. Zhurnal Pikladnoi Khimii. 1977;**50**:1181

[27] Malik WU, Srivastava SK, Kumar S. Ion-exchange behaviour of pyridinium tungstoarsenate. Talanta. 1976;**23**:323

[28] Chetverina RB, Baichinova ES. Ibid. 1977;**50**:1183

[29] Niwas R, Khan AA, Varshney KG. Synthesis and ion exchange behaviour of polyaniline Sn(IV) arsenophosphate: A polymeric inorganic ion exchanger. Colloids and Surfaces, A: Physiochemical and Engineering Aspects. 1999;**150**:7

[30] Varshney KG, Pandith AH. Chemical and Environmental Research. 1996;**5**(1-4):141

[31] Sanchez C, Ribot F. Design of hybrid organic-inorganic materials synthesized via sol-gel chemistry. New Journal of Chemistry. 1994;**18**:1007

[32] Varshney KG, Jain V, Tayal N. Synthesis and ion-exchange behaviour of acrylamide zirconium(IV) phosphate: A novel crystalline and Hg(II) selective hybrid inorganic ion exchanger. Indian Journal of Chemical Technology. 2003;**10**:186

[33] Varshney KG, Jain V, Agrawal A, Mojumdar SC. Pyridine based zirconium(IV) and tin(IV) phosphates as new and novel intercalated ion exchangers. Journal of Thermal Analysis and Calorimetry. 2006;**86**:3,609

[34] Varshney KG, Gupta P. Synthesis and characterization of a mercury selective phase of acrylamide tin(IV) phosphate hybrid ion exchanger: Separation of Hg(II) from Cd(II), Pb(II) and Sr(II). Indian Journal of Chemistry. 2003;**42A**:2974

[35] Varshney KG, Pandith AH. Synthesis and ion exchange behavior of acrylonitrile-based zirconium phosphate - A new hybrid cation exchanger. Journal of the Indian Chemical Society. 2001;**78**:250

[36] Varshney KG, Tayal N, Khan AA, Niwas R. Synthesis, characterization and analytical applications of lead (II) selective polyacrylonitrile thorium (IV) phosphate: A novel fibrous ion exchanger. Colloids and Surfaces A: Physicochemical and Engineering Aspects. 2001;**181**:123

[37] Varshney KG, Tayal N, Gupta P, Agrawal A, Drabik M. Synthesis, ion-exchange and physico-chemical studies on a polystyrene cerium(IV) phosphate hybrid fibrous ion exchanger. Indian Journal of Chemistry. 2004;**43A**:2586

[38] Varshney KG, Tayal N. Polystyrene thorium(IV) phosphate as a new crystalline and cadmium selective fibrous ion exchanger. Synthesis characterization and analytical applications. Langmuir. 2001;**17**:2589

[39] Varshney KG, Tayal N, Gupta U. Acrylonitrile based cerium (IV) phosphate as a new mercury selective fibrous ion-exchanger. Colloids and Surfaces A: Physico-chemical and Engineering Aspects. 1998;**145**:71

[40] Varshney KG, Gupta P, Tayal N. Synthesis, characterization and applications of a new phase of crystalline and mercury selective acrylamide cerium(IV) phosphate: A novel fibrous ion exchanger. Indian Journal of Chemistry. 2003;**42A**:89

[41] Mojumdar SC, Varshney KG, Agrawal A. Hybrid fibrous ion exchange materials: Past, present and future. Research Journal of Chemistry and Environment. 2006;**10**(2):85

[42] Varshney KG, Agrawal A, Mojumdar SC. Pectin based cerium (IV) and thorium (IV) phosphates as novel hybrid fibrous ion exchangers synthesis, characterization and thermal behaviour. Journal of Thermal Analysis and Calorimetry. 2005;**81**:183

[43] Varshney KG, Drabik M, Agrawal A. Cellulose acetate based thorium(IV) phosphate as a new and novel hybrid fibrous cation exchanger: Synthesis, characterization and thermal analysis. Indian Journal of Chemistry. 2006;**45A**:2045

[44] Varshney KG, Agrawal A, Mojumdar SC. Pyridine based cerium(IV) phosphate hybrid fibrous ion exchanger. Journal of Thermal Analysis and Calorimetry. 2007;**90**(3):731

[45] Varshney KG, Agrawal A, Mojumdar SC. Pyridine based thorium(IV) phosphate hybrid fibrous ion exchanger. Journal of Thermal Analysis and Calorimetry. 2007;**90**(3):721

[46] Varshney KG, Rafiquee MZA, Somya A, Drabik M. Synthesis and characterization of a Hg(II) selective n-butyl acetate cerium(IV) phosphate as a new intercalated fibrous ion exchanger: Effect of surfactants on the adsorption behaviour. Indian Journal of Chemistry. 2006;**45A**:1856

[47] Tanford C. The Hydrophobic Effect: Formation of Micelles and Biological Membranes. 2nd ed. New York: Wiley; 1980

[48] Rosen MJ. Surfactants and Interfacial Phenomena. New York: Wiley; 1978

[49] Cullum DC. Introduction of Surfactant Analysis. 1st ed. UK: Chapman and Hall; 1994

[50] Linfield WM, editor. Anionic Surfactants. New York: Dekker M; 1973

[51] Jungermann E. Dekker M. Cationic Surfactants. New York; 1970

[52] Schick MJ, editor. Nonionic Surfactants. New York: Dekker M; 1967

[53] Porter MR. Handbook of Surfactants. 2nd ed. UK: Chapman and Hall; 1994

[54] Menger FM, Littau CA. Gemini surfactants: A new class of self-assembling molecules. Journal of the American Chemical Society. 1993;**115**:10083

[55] Mukerjee P, Banerjee K. A study of the surface pH of micelles using solubilized indicator dyes. The Journal of Physical Chemistry. 1964;**68**:3567

[56] Preston WC. Some correlating principles of detergent action. The Journal of Physical and Colloid Chemistry. 1948;**52**:84

[57] Shinoda K, Nakagawa T, Tamamushi BI, Isemura T. Colloidal Surfactants, Some Physico-Chemical Properties. New York: Academia Press; 1963

[58] Mukerjee P, Mysels KJ. Critical Micelle Concentrations of Aqueous Surfactant Systems. Washington, DC: National Bureau of Standards, NSRDS-NBS 36; 1971

[59] Dill KA, Flory P. Molecular organization in micelles and vesicles. Proceedings of the National Academy of Sciences. 1981;**78**:676

[60] Luisi PL, Straub B. Reverse Micelles—Technological and Biological Relevance, Plenum Press New York; 1984:143

[61] Harwell JH, Haskins IC, Schecter RS, Wad WH. Langmuir. 1985;**1**:251

[62] Rosen MJ. Surfactants and Interfacial Phenomena. 2nd ed. New York: Wiley; 1989

[63] Hearn MT, editor. Ion-Pair Chromatography. New York: Dekker M; 1985. p. 251

[64] Hoffmann H, Ebert G. Surfactants, micelles and fascinating phenomena. Angewandte Chemie. 1988;**100**:933

[65] Rosen MJ. Surfactants and Interfacial phenomenon. 2nd ed. New York: Wiley; 1989

[66] Myers D. Surfactnat Science and Technology, VCH New York; 1988: 351

[67] Karsa DR, editor. Industrial Applications of Surfactants. Cambridge, UK: Royal Society of Chemistry; 1987

[68] Karsa DR. Industrial Applications of Surfactants. 2nd ed. Heidelberg, Germany: Springer-Verlag; 1990

[69] Karsa DR. Industrial Applications of Surfactants. 3rd ed. Heidelberg, Germany: Springer-Verlag; 1992

[70] Karsa DR. Industrial Applications of Surfactants. 4th ed. Cambridge, UK: Royal Society of Chemistry; 1999

[71] Varshney KG, Rafiquee MZA, Somya A. Effect of surfactants on the adsorption behaviour of cerium (IV) phosphate, cation exchanger for alkaline earths and heavy metal ions. Colloids and Surfaces A: Physicochemical and Engineering Aspects. 2007;**301**:69

[72] Varshney KG, Rafiquee MZA, Somya A. Effect of surfactants on the adsorption behaviour of tin (IV) phosphate, cation exchanger for alkaline earths and heavy metal ions. Colloids and Surfaces A: Physicochemical and Engineering Aspects. 2007;**301**:224

[73] Somya A, Upadhyaya V. Synthesis, characterization and adsorption behaviour of N-dodecyl pyridinium chloride based Tin (IV) phosphate: A new surfactant based hybrid ion exchanger. Research Journal of Chemistry and Environment. 2017;**21**(1):18

[74] Somya A, Rafiquee MZA, Varshney KG, Al-Lohedana HA. Synthesis, characterization and analytical applications of N-dodecyl pyridinium chloridecerium (IV) phosphate fibrous ion exchanger: Selective for Hg(II) and its binary separations. Desalination and Water Treatment. 2015;**56**:638

[75] Somya A, Rafiquee MZA, Varshney KG. Synthesis, characterization and analytical applications of sodium dodecyl sulphate cerium (IV) phosphate: A new Pb (II) selective, surfactant-based intercalated fibrous ion exchanger. Colloids and Surfaces A: Physicochemical and Engineering Aspects. 2009;**336**:142

[76] Varshney KG, Rafiquee MZA, Somya A. Triton X-100 based cerium(IV) phosphate as a new Hg(II) selective, surfactant based fibrous ion exchanger: Synthesis, characterization and adsorption behaviour. Colloids and Surfaces A: Physicochemical and Engineering Aspects. 2008;**317**:400

[77] Varshney KG, Rafiquee MZA, Somya A. Synthesis, characterization and adsorption behaviour of TX-100

based Sn(IV) phosphate, a new hybrid ion exchanger. Journal of Thermal Analysis and Calorimetry. 2007;**90**:663

[78] Iqbal N, Rafiquee MZA. Synthesis and characterization of lead (II) selective sodium dodecyl benzene sulphonate-cerium (IV) phosphate ion exchanger. Colloids and Surfaces A: Physicochemical and Engineering Aspects. 2010;**364**:67

[79] Iqbal N, Mobin M, Rafiquee MZA, Al-Lohedana Hamad A. Removal of Cu2+, Cd2+, Pb2+ and Hg2+ ions by the synthesized sodium dodecyl benzene sulphonate-based tin (IV) phosphate (SDBS-SnP) cation exchanger. Desalination and Water Treatment. 2013;**51**:6688

[80] Iqbal N, Mobin M, Rafiquee MZA. Synthesis and characterization of sodium bis (2-ethylhexyl) sulfosuccinate based tin (IV) phosphate cation exchanger: Selective for Cd2+, Zn2+ and Hg2+ ions. Chemical Engineering Journal. 2011;**169**:43

[81] Iqbal N, Mobin M, Rafiquee MZA, Al-Lohedana Hamad A. Characterization and adsorption behavior of newly synthesized sodium bis(2-ethylhexyl) sulfosuccinate–cerium (IV) phosphate (AOT–CeP) cation exchanger. Chemical Engineering Research and Design. 2012;**90**:2364

Section 2

Flexible Electronics

Chapter 5

Pressure-Sensitive Adhesives for Flexible Display Applications

Tae-Hyung Lee, Ji-Soo Kim, Jung-Hun Lee and Hyun-Joong Kim

Abstract

Pressure-sensitive adhesives (PSA) have been used in electronics for not only attachment of two materials but also shock absorption, thermal and electrical conductivity, electromagnetic shielding, and optical property. Optically clear adhesives (OCA) have been used as a core material for optical performance of display. In addition to basic properties of OCA such as adhesion strength, transmittance, haze, and reliability, it has required dielectric constant, gap filling, and anticorrosion according to a substrate. However, the structural and functional changes of flexible display bring challenges to OCA that protects vulnerable components such as thin-film transistor, OLED, and thin-film encapsulation by stress dispersion and adjustment of a neutral plane. At the same time, flexibility and existing properties are essential. In this chapter, the development of components and performance of OCA, and evaluation methods will be discussed.

Keywords: pressure-sensitive adhesives, PSA, optically clear adhesives, OCA, display, flexible, foldable

1. Introduction

PSA is a material attached by slight pressure and detached easily from the substrate [1]. PSA's semisolid property caused by low glass transition temperature (T_g, generally 25–45°C below usage temperature) at usage temperature makes these features possible [2].

PSA can be classified by type and proportion of components into natural and synthetic rubber, acrylic, silicone, and urethane PSA [3]. The acrylic PSA is widely used in display, mobile phones, and automotive applications due to its high transparency, weather resistance, heat resistance, and high adhesion strength [4–6]. Furthermore, the acrylic PSA can have a wide range of properties because of a variety of acrylic monomers. Although the silicone PSA is not as universal as the acrylic PSA, special applications have been used that require high reliability because of its high resistance to high and low temperatures.

PSA has better processability than liquid-type adhesive because it can adhere without the hardening process and to the three-dimensional substrate. Also in the automobile industry, PSA has been considered as a substitutive process of the traditional mechanical joining method using bolts and screws because of the requirement of weight reduction. In particular, as the application of electronic and automobile industries is expanded, it is required to have not only adhesion strength and processability but also complex functions such as electrical and thermal conductivity, high thermal resistance, and reliability of humidity or chemicals.

OCA is used for attachment of layers in the display including cover window, touch panel, polarizers, and the light-emitting layer, which commonly require high transmittance, low haze, and corrosion resistance to ITO (indium tin oxide) film. When OCA is directly bonded to ITO film, the acid component must be excluded for the durability of the ITO film. Prolonged contact with acid can cause touch problems by reacting with metal, which makes surface resistance to increase [7]. It is also important for OCA to minimize air bubbles to reduce defects.

The demand of thin and flexible displays is increasing as interest in small and diverse designs grows [8]. Flexible displays can be distinguished by their intended use and function [9]. Recently, significant progress has been made in achieving active-matrix organic light-emitting diodes with bendable and rollable displays [10–12]. Several PSA properties are additionally required to create these flexible displays. Generally, OCA for flexible displays requires low shear modulus and T_g [13]. The recovery and stress relaxation properties of OCA are important for flexible displays because a high recovery of PSA prevents it from deformation under repeated folding-unfolding conditions [14].

2. Component of PSA

As mentioned above, acrylic PSA has been widely used for not only household or general industrial product but also electronics and automobile industries due to various advantages such as adhesion properties, optical properties, high reliability, and easy modification. Especially in OCA manufacturing, the synthesis and curing by ultraviolet (UV) radiation are used for short process time, high molecular weight, and nonsolvent process. The acrylate is one of the most suitable materials for the UV process. In this chapter, we would like to describe the materials that compose acrylic PSA including OCA.

acrylic PSA is synthesized by selecting several acrylic monomers, and the monomers can be classified into the alkyl (meth)acrylate and functional (meth) acrylate. Also, the alkyl (meth)acrylate is divided into monomers with a low T_g and a high T_g. The acrylic monomer of low T_g has a linear carbon chain that consists of 4–17 atoms, and the monomer of high T_g has a short linear chain (1–3 carbon atoms) or bulky chemical structure such as a cyclic hydrocarbon or aromatic ring. The functional monomer has a hydrophilic functional group such as a carboxyl group and hydroxyl group, and this gives an acrylic polymer reaction site for crosslinking and hydrophilicity to enhance adhesion and cohesion strength. The low T_g monomers are the main component of acrylic PSA, and the high T_g monomers and the functional monomers are added to adjust the characteristics of acrylic PSA. A typical acrylic PSA consists of 50–90% of the low T_g monomer, 10–40% of the high T_g monomer, and 2–20% of the functional monomer [3, 15].

The adhesion properties of PSA are determined by both the viscoelasticity and chemical characteristics. The relationship between the viscoelasticity and the adhesion has been studied, and the range of storage modulus (G′) and loss modulus (G″) is suggested as a viscoelastic window [3, 16–19]. The surface free energy can represent the chemical characteristic of PSA. Wettability between PSA and substrate is determined by the relation of surface free energy of PSA and substrate. The immediate adhesion by quick wetting should be demonstrated right after contact with the substrate for effective use of PSA [20]. Kowalski et al. conducted a study about the tack value of acrylic PSA that increased from 300 to 700% depending on the cross-linking degree when surface free energy of a polymer substrate increased from 20.5 to 42.9 mJ/m^2 [21].

Silicone PSA has extreme resistance at a wide temperature range from −40 to 300°C. In addition, because it can be applied to substrates having various surface energies, it is used for masking tape for printed circuit board plating, electrical insulation tape, and OCA. However, due to its drawbacks such as high price, high process temperature, and high release strength to release film, it has been applied only to specific fields that can withstand harsh environments [22]. This PSA consists of silanol-terminated silicone polymer and silanol-functional siloxane resin. The silicone polymer for PSA is a semisolid gum having a high viscosity and 600–1000 kDa of molecular mass [23].

Many manufacturers of OCA adopt the UV radiation process for the mentioned reasons, and the process is divided into two steps. The first step is a synthesis of an acrylic prepolymer by UV radiation to a mixture of the monomers and photoinitiator. The monomer mixture, except for the photoinitiator, is similar to the composition described above. But, because the acrylic acid, the major functional monomer, cannot be used, other functional monomers are evaluated to enhance adhesion strength. N-vinyl caprolactam (NVC) and N, N-dimethyl acrylamide (DMAA) are the representative monomers [24, 25]. Since nitrogen atom has a high electronegativity, it can improve the cohesion and at the same time prevent the risk of corrosion by acid [26]. In the second step, the synthesized acrylic prepolymer is mixed with multifunctional acrylate as a cross-linking agent and cross-linked by UV radiation. However, the unreacted monomers can be remained after the curing process. Because a drying process is not included in the UV radiation process, the unreacted monomers can be evaporated in the final product [27].

Currently, the development issue facing OCA is flexibility, and this not only prohibits the use of acidic monomers but also limits the selection of other common monomers. The high T_g monomers and functional monomers cause high modulus, and this property restricts the deformation of OCA. 2-Ethylhexyl acrylate (2-EHA) and butyl acrylate (BA) are used as a conventional low T_g monomer, but other monomers with longer carbon chains are considered for flexibility.

3. Development in display and OCA

All kinds of displays have a multilayer structure, and most of these layers are combined with OCA or PSA (**Figure 1**). Without OCA, the light from the backlight

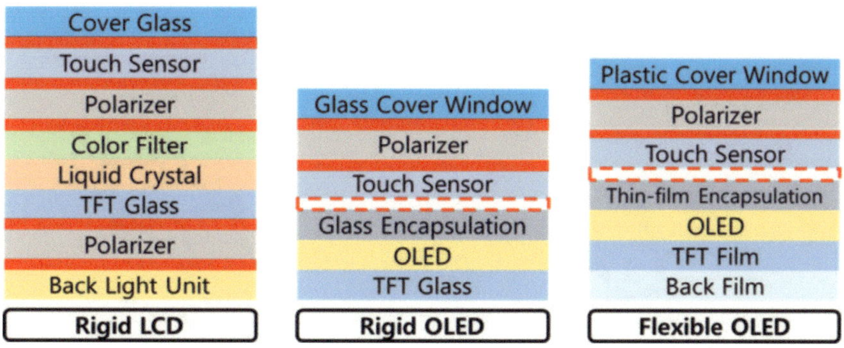

Figure 1.
Multilayer structures of the display. OCA or PSA can be applied to red layers in the structures. A touch sensor can be placed on the encapsulation by OCA or direct deposition.

unit is reflected at the interface between each film and the air due to the differences of the refractive index. Finally, less than 10% of light reach to user's eyes [28]. The presence of OCA affects the clarity of the display screen because OCA between films not only holds the film together but also prevents loss of light. Also, the difference in the refractive index between layers decreases, and a similar refractive index allows the straight progress of light without loss [29]. So, OCA should basically be transparent and have a low haze. It should also be optically isotropic, with less coloring and discoloration in environmental conditions [30].

LCD is the most widely used display to date. As shown in **Figure 1**, because it has to have a more complex structure than OLED display, more PSA layers are used. While the polarizer is used for blocking reflection of external light in OLED, it plays an important role in displaying images in LCD. The most important thing for the polarizer in LCD is to minimize the light leaking [31–34]. According to the study of Ma et al., OCA for polarizer can improve light leaking by reducing the shrinkage stress between a glass substrate and polarizer [35].

OLED is based on pixels that emit light on their own, and it can display pure black by turning off the pixels. The structures between LCD and OLED show stark differences because of their self light-emitting layer and vulnerability to oxygen and moisture [36]. Although there are many technical differences between OLED and LCD, the requirements for OCA are not much different.

A key to next-generation display technology is flexibility. The flexible displays are divided into quasi-flexible and real-flexible. The quasi-flexible display contains curved and bended displays, and Samsung's Galaxy Round is the first smartphone with a curved display. The manufacturing process of the curved display includes the bending process of the OLED panel. When the display panel is bent, spring back force is generated in the plastic film of the module to reverse the elastic deformation [37]. So, OCA of the curved module should have high adhesion property to resist this force. Otherwise, there is a risk of creating air bubbles when OCA is separated away from the glass. In order to achieve high adhesion and filling thick ink step at

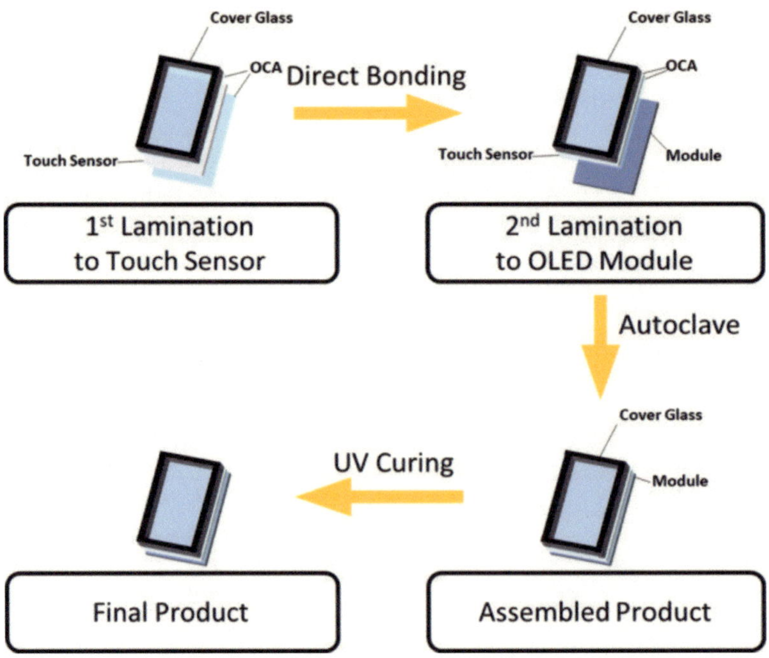

Figure 2.
Application process of the 3 M CEF3806.

the same time, 3 M has developed OCA products that have low modulus and good wetting performance with semicured structure (CEF3806, 3 M). Semicured OCA shows optimal adhesion performance through additional UV curing after lamination as shown in **Figure 2**.

The foldable display is a united panel containing a variety of layers and substrates. For a flexible OLED display, it is desirable to have either an OCA that perfectly mechanically decouples the layers from each other in bending (i.e., a material with no modulus) to minimize strain on critical layers or to have OCA that can tune the position of the neutral plane during the folding process [38]. Also, although the thickness of OCA is thinner than that of OCA for a rigid display for reducing strain caused by folding, the strong adhesion, flexibility, and durability to withstand hundreds of thousands of bends are required.

4. Stress dispersion and neutral plane

OLED type of display has been chosen to implement flexible display. Major components of OLED display are thin-film transistors (TFTs), organic light-emitting diodes (OLEDs), and thin-film encapsulation (TFE). However, these layers are easily damaged by minor deformations and lose their functions because they are fragile in common [39]. To avoid this problem, the materials that have low modulus are used to absorb stress, and structural design to place these parts in the neutral plane is applied [40]. When the display composed with multilayers is bent, tensile or compressive deformation occurs depending on the position, but for each design, the layer at a particular position is simply bent without such deformation. This position is called the neutral plane. Masumi Nishimura *et al*. give the following equation for calculating the position of the neutral plane [41]:

$$\lambda = \sum_{i=1}^{n} E_i\left(h_i^2 - h_{i-1}^2\right)/2\sum_{i=1}^{n} E_i t_i \quad (1)$$

The λ means the distance from the surface of the innermost layer of the multilayer to the neutral plane, while E_i, h_i, and t_i mean the modulus of i-th layer, the distance from the innermost surface to the layer, and the thickness of the layer, respectively. As expressed in the equation, the position of neutral plane is determined by the modulus and thickness of each layer. Because OCA layers are relatively easy to control modulus and thickness without losing function, OCA layers are able to be used for adjustment of the location of the neutral plane.

The studies were conducted to protect several layers by splitting neutral planes within a multilayer. Su et al. utilized PSA layers in flexible piezoelectric mechanical energy harvesters to split the neutral planes, thereby implementing a flexible structure without destroying the components vulnerable to deformation [42]. Based on this research, Nishimura et al. published a study about splitting neutral plane using OCA in a foldable AMOLED structure. The trend of neutral plane formation according to the modulus of adhesive was identified, and the splitting was observed by the use of low modulus adhesive in the system [41].

5. Test methods for foldable OCA

Recently, foldable mobile phone, which is a type of flexible electronics, has been commercialized. So, manufacturers of mobile devices, display, and materials for electronic devices and research institutes have been set evaluation methods for foldable characteristics. The most representative and essential method is the folding test. Samsung Electronics, a major manufacturer of electronics including mobile phones, released folding test images with the launch of Galaxy Fold in September 2019 [43]. Folding durability not only of the assembled mobile phones but also of single or laminated film used as part of electronics has been evaluated. This test is performed with various environments and test modes such as low or high temperature, high humidity, the radius of curvature, static fold, and dynamic fold for simulation of various usage environments [44].

The folding test can obtain results from a complex interaction between multilayered components. So, the following various methods have been conducted to evaluate OCA independently.

5.1 Shear strain on folded OCA

When a book is bent, it can be seen that the unbound edge is deformed into an inclined form. Because unlike the book, however, each layer of the foldable display is attached to each other by PSA, each layer of multilayered display is deformed by different type and extent of stress. In this case, as shown in **Figure 3**, the outer layers and inner layers of the neutral plane are subjected to tensile stress and compressive stress, respectively. As such, the stress applied to each layer during folding can be calculated by the following Equation [41]:

$$\sigma = E_i \varepsilon = E\,i(y - \lambda)/\rho \tag{2}$$

Using a simple equation to calculate the stress (σ) by modulus (E_i) and strain (ε), the stress can be obtained by the positional difference between the target layer and the neutral plane ($y - \lambda$) and the folding radius (ρ) when the modulus of the material is fixed. Since these deviations of strain occur not only between layers but also within a single layer as shown in **Figure 4**, each layer including OCA is subjected to shear stress. So, most tests for evaluating foldable OCA are conducted under shear stress.

Although the shear strain depends on the thickness and placement of other layers, the shear strain of >300% is generally applied to the adhesive layer at a 5 mm radius of curvature. If the radius was reduced to 3 mm, the shear strain is increased up to 500–700% [45]. For the reduction of increase of shear strain by change of radius, the minimization of thickness and optimization of the structural design of multilayers have been studied [46].

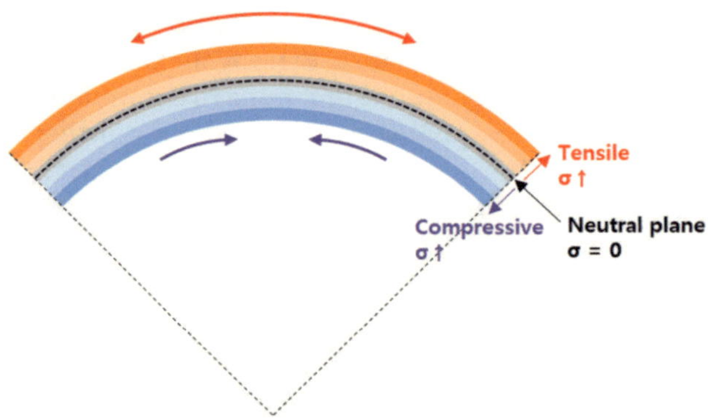

Figure 3.
Schematic diagram of applied stresses of folded multilayered display including neutral plane [41].

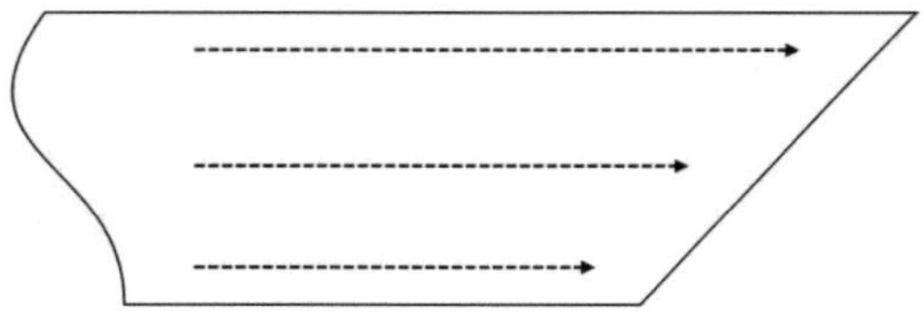

Figure 4.
Schematic diagram of shear deformation of a single layer.

5.2 Adhesion strength

Although the adhesion test is not a specific method for OCA applied to the foldable display, it is an important and fundamental method to evaluate all kinds of PSA. The test mode is selected according to the external force applied between PSA and substrate, and the general methods are peel, tack, lap shear, and pull-off test. The 180° peel test and lap shear test are conducted for foldable OCA (**Figure 5**).

Although the evaluation of adhesion is not a specific test for foldable OCA, this is considered an important factor because of several limits for satisfying adhesion strength. One of these limits is the substitution of the glass cover window to the plastic cover window for foldable properties [47]. The surface energies of several substrates such as plastic, glass, and SUS are listed in **Table 1** [48, 49]. The polyimide (PI) film has been used for the plastic cover window, and the surface energy of plastic substrate including PI is significantly lower than the surface energy of glass and SUS substrates. According to the rule of thumb, the lower surface energy of a substrate is able to decrease adhesion strength between PSA and substrate [50]. It is necessary to increase adhesion strength to the plastic substrate with minimization of rising T_g and modulus of OCA. If sufficient adhesion strength is not developed, it will not be possible to withstand the stress applied between OCA and substrate in continuous folding. The other is the limitation of the usable monomers for increasing adhesion strength as described above.

5.3 Shear modulus

The representative methods to measure shear modulus of PSA are the shear sandwich method by Dynamic Mechanical Analyzer (DMA) and the torsion method by Advanced Rheometric Expansion System (ARES) (**Figure 6**). Shear sandwich and torsion methods are subject to uniaxial and rotational shear stresses, respectively. In both of these methods, the material responds to repetitive deformation, which has a constant frequency, and its viscoelastic property is evaluated by this reaction. The elastic property reacted immediately is represented by the storage modulus, and the viscous property reacted belatedly is represented by the loss modulus [3]. Generally, the shear modulus is the storage modulus and is presented G' separated with the tensile modulus (E'). Since the shear sandwich method can have inaccuracy of pressure in the loading sample, the torsion method has been preferred.

When measuring the shear modulus of foldable OCA, Campbell et al. suggest that the storage modulus should be maintained and has a low T_g in the operating range (−20 to 80°C) [38]. Because, as presented in **Figure 7**, T_g is shown at the primary reduction of storage modulus, lowering T_g can result in modulus reduction below the operating range. Since the temperature affects the movement of molecules, polymers recognize the time or frequency as a different degree

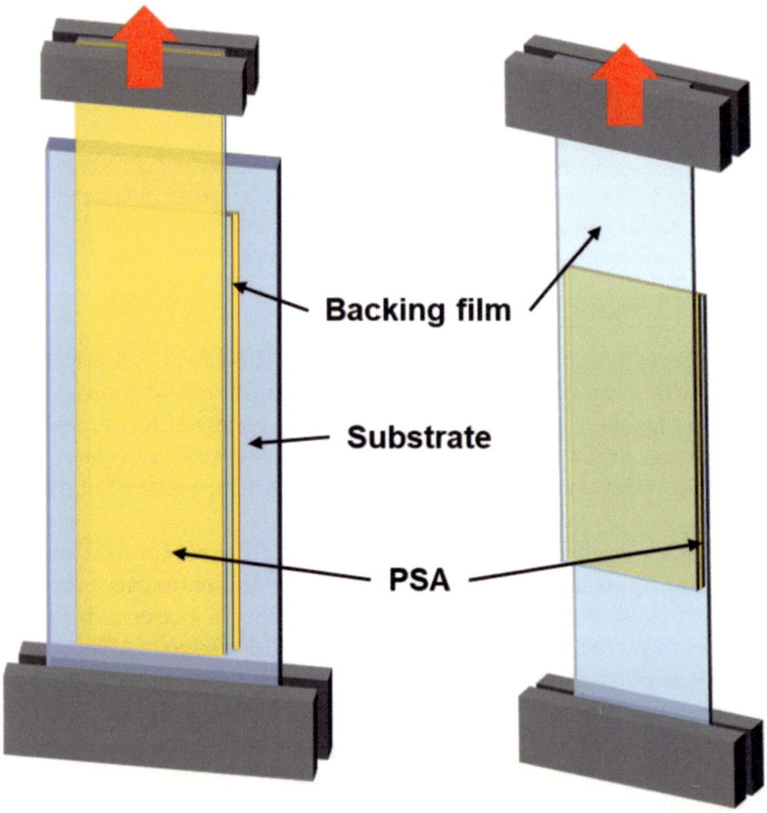

Figure 5.
180° peel and lap shear test for adhesion evaluation.

Substrate	Polypropylene	Polyester	Polyimide	Glass	SUS
Surface energy (mN/m)	29	43	50	200–300	700–1100

Table 1.
Surface energy of substrates.

according to temperature [51]. Therefore, reducing modulus at low temperatures maintains the flexibility not only in low usage temperature but also in rapid deformation.

The maintenance of the shear modulus ensures adhesion strength [38], strain recovery, and processability at high temperature. If the modulus decreased as the temperature rise, the adhesion strength and strain recovery are able to decrease by cohesion reduction.

5.4 Shear creep and stress relaxation

The creep is the strain change with time under constant stress, and the stress relaxation is the stress change with time under a constant strain. Furthermore, the strain recovery after removing stress is an important factor to assess the restoration of PSA after folding. By the folding test, only the functional and macroscopic properties occurred after repetitive or static deformation can be observed, so the flexibility and resilience of a single material can be evaluated through creep and stress relaxation.

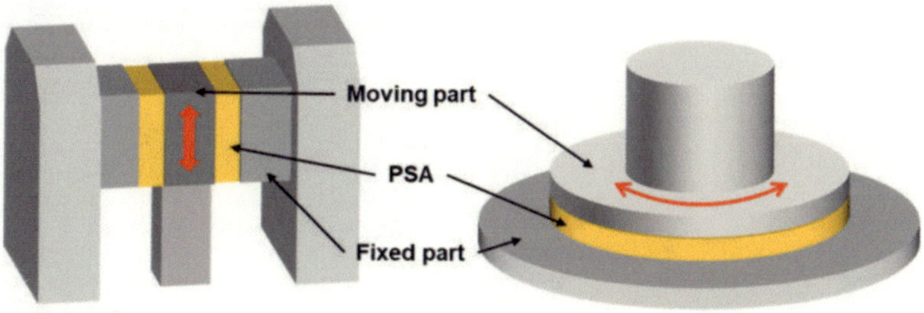

Figure 6.
Shear sandwich and torsion methods for testing shear modulus.

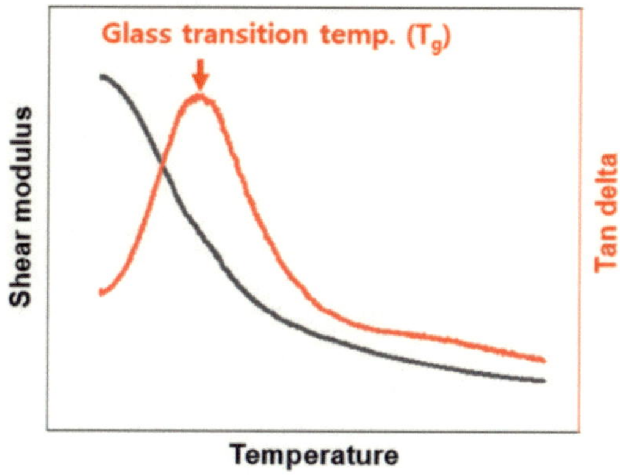

Figure 7.
The graph of shear modulus and tan delta.

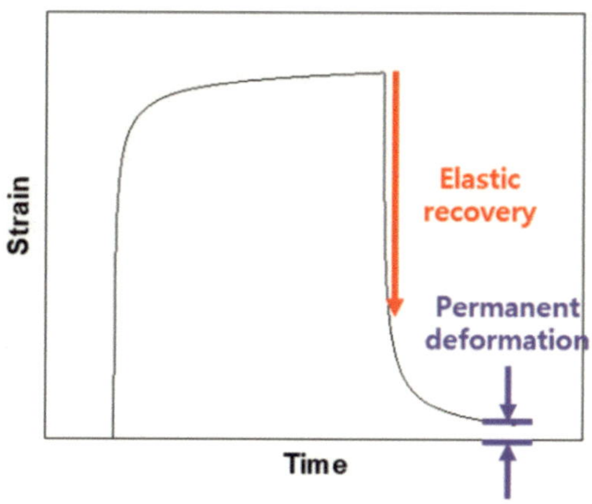

Figure 8.
The creep behavior of OCA.

In the creep test, after immediate deformation of OCA by constant stress, the strain increases continuously because of its viscoelasticity. The elastic property makes the strain to be maintained after the initial deformation, but the viscous property results in an increase of continuous deformation. Lee et al. evaluated creep recovery by dividing it into elastic recovery and residual strain after a specific time (**Figure 8**) [52]. While the elastic deformation is recovered after the removal of stress, the viscous deformation causes permanent deformation that cannot be recovered [53].

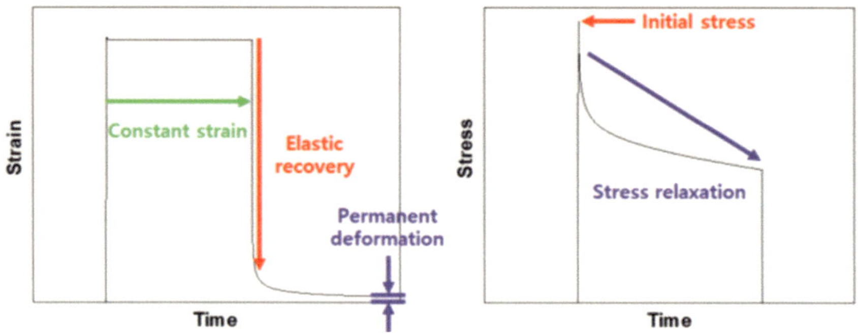

Figure 9.
The stress relaxation behavior of OCA.

In the stress relaxation test, the continued decrease of stress is observed after the highest stress appears in response to the initial rapid deformation (**Figure 9**). This reduction is due to molecular movement, bond break, and bond interchange [51]. Energy reduction due to the stretching of molecules and intermolecular motion within a range fixed by cross-linking or entanglement is mostly recoverable after stress removal, but the breakage or interchange of the intermolecular bond and intermolecular slip is irreversible.

Permanent deformation that cannot be recovered after the removal of stress in creep and stress relaxation measurement can cause wrinkles or optical defects during the folding test of the display module or final product. Considering the viscoelasticity of PSA and interaction with adjacent films that have higher modulus, the perfect recovery is not required for OCA. But, high resilience as possible may be advantageous in long-term reliability.

6. Researches of OCA for flexible display

Lee et al. researched several factors that affect the foldable properties of OCA. First is the molecular weight of the prepolymer. The acrylic resins with a molecular weight in the rage of 360,000–690,000 are synthesized, and stress relaxation and recovery properties are evaluated under a constant shear strain of 400%. Although the initial stress of the stress relaxation test increases as the molecular weight increases, the ratio of stress relaxation decreases. The recovery after stress relaxation and creep is also improved [52]. Further, as the content of the cross-linking agent increased, the recovery increases in stress relaxation evaluation. But, it is confirmed that it decreases at over specific content. As the content of the cross-linking agent increases, the cross-link point increases. And the viscous deformation decreases. But, since too many cross-link points restrict the deformation, it results in bond breakage in the strain of 400% and decrease recovery [54]. As such, the entanglement and chemical connection between molecules play a major role in deformation and recovery.

They reported adhesion properties and recovery behavior of cross-linked silicone OCA using a platinum catalyst. The effect of the degree of cross-link is similar to that of acrylic OCA. But, the speed and degree of elastic recovery after stress relaxation is much faster and better than acrylic OCA [55]. Lee et al. also attempted to improve adhesion properties and recovery by connecting a styrene-isoprene-styrene (SIS) elastomer to acrylic OCA. The entanglement of the elastomer shows a positive effect on adhesion and recovery. However, the styrene groups of SIS elastomer cause the low transparency of OCA [56].

Campbell et al. present the shear modulus of standard OCA, foldable OCA, and improved foldable OCA, and results of repetitive tension tests. Compared to standard OCA, the foldable OCA's low modulus is suitable for foldable equipment, and the improved foldable OCA has the modulus of 10–100 kPa even at high temperatures by preventing modulus decrease at high temperature. The decrease of modulus causes poor adhesion and mechanical durability in high temperature. So, they are researching that the foldable OCA has a stable modulus while maintaining a low T_g [38].

Some studies conduct a simulation method to figure out the roles of OCA. Salmon et al. demonstrate the effect of the modulus and elasticity of OCA by modeling of foldable OLED panels, and two types of OCA are used. The first one is 3 M foldable OCA and has lower modulus and higher elasticity, and the other one is 3 M OCA 8180. The softer OCA, 3 M foldable OCA, contributes less bending stress of display panel and less tensile strain of OLED layer than the other OCA. Also, they show buckling due to residual strain after unfolding. The elasticity of 3 M foldable OCA prevents continuous shear deformation and shear creep, and results in lower buckling than 3 M OCA 8180 [57]. Jia et al. presented nonlinear viscoelastic behavior of OCA in a folded display by constitutive model. They show that each layer in a multilayered display has its own neutral plane when OCA exists between the layers and this phenomenon due to the decoupling effect of the soft OCA. Also, they conduct the folding simulation of panel structures with various thicknesses of OCA. The major factor that affects the strain of the first OCA layer laid under a cover window is its own thickness. Consequentially, using thicker first OCA layer improves thermal foldability, because it reduces its own strain and risk of delamination [58].

7. Summary

In the flexible display, OCA is used for protecting TFT, TFE, and OLED in addition to the existing roles in the flat display. The low T_g and fixed modulus at operating temperature range are required to foldable OCA, while it secures high adhesion strength to the plastic substrate. Some high T_g and functional monomers used for the increase of adhesion strength have a limit to be used because of its high T_g and modulus. Also, the low T_g monomers that have long side chains are considered for OCA's high elongation and recovery. The foldable durability can be evaluated by the creep and stress relaxation test, and the strain and recovery of OCA are a combination of the elastic and viscous part. The viscous part can cause an optical defect in the display by permanent deformation. Because the perfect recovery of OCA, viscoelastic material with log T_g, and modulus is difficult, studies to compensate permanent deformation by the interaction with adjacent layers, structural design, and control of component have been conducted.

Author details

Tae-Hyung Lee, Ji-Soo Kim, Jung-Hun Lee and Hyun-Joong Kim*
Laboratory of Adhesion and Bio-Composites, Program in Environmental Materials Science, College of Agriculture and Life Sciences, Seoul National University, Seoul, Republic of Korea

*Address all correspondence to: hjokim@snu.ac.kr

References

[1] Webster I. Recent developments in pressure-sensitive adhesives for medical applications. International Journal of Adhesion and Adhesives. 1997;**17**(1):69-73. DOI: 10.1016/S0143-7496(96)00024-3

[2] Creton C. Pressure-sensitive adhesives: An introductory course. MRS Bulletin. 2003;**28**(6):434-439. DOI: 10.1557/mrs2003.124

[3] Satas D. Handbook of Pressure Sensitive Adhesive Technology. 3rd ed. Vol. 171-202. Warwick, Rhode Island (USA): Satas & Associates; 1999. pp. 444-450

[4] Gower M-D, Shanks R-A. The effect of varied monomer composition on adhesive performance and peeling master curves for acrylic pressure-sensitive adhesives. Journal of Applied Polymer Science. 2004;**93**(6):2909-2917. DOI: 10.1002/app.20873

[5] Benedek I. Development and Manufacture of Pressure-Sensitive Products. United States: CRC Press; 1998

[6] Chun H, Kim H-A, Kim G, Kim J, Lim K-Y. Effect of the stress relaxation property of acrylic pressure-sensitive adhesive on light-leakage phenomenon of polarizer in liquid crystal display. Journal of Applied Polymer Science. 2007;**106**(4):2746-2752. DOI: 10.1002/app.26889

[7] Park J-W, Yoo T-M, Chung H-W, Jang W-B, Kim H-J. Evaluation of UV-curability of photo-curable materials for acid free pressure sensitive adhesive. Journal of Photopolymer Science and Technology. 2011;**24**(5):553-560. DOI: 10.2494/photopolymer.24.553

[8] Kim S, Choi W, Rim W, Chun Y, Shim H, Kwon H, et al. A highly sensitive capacitive touch sensor integrated on a thin-film-encapsulated active-matrix OLED for ultrathin displays. IEEE Transactions on Electron Devices. 2011;**58**(10):3609-3615. DOI: 10.1109/TED.2011.2162844

[9] Crawford GP. Flexible flat panel display technology. West Sussex: Wiley; 2005:1-4. DOI: 10.1002/0470870508

[10] Jin D-U, Kim T-W, Koo H-W, Stryakhilev D, Kim H-S, Seo S-J, et al. 47.1: Invited paper: Highly robust flexible AMOLED display on plastic substrate with new structure. SID Symposium Digest of Technical Papers. 2010;**41**(1):703-705. DOI: 10.1889/1.3500565

[11] An S, Lee J, Kim Y, Kim T, Jin D, Min H, et al. 47.2: 2.8-inch WQVGA flexible AMOLED using high performance low temperature Polysilicon TFT on plastic substrates. SID Symposium Digest of Technical Papers. 2010;**41**(1):706-709. DOI: 10.1889/1.3500566

[12] Kim S, Kwon H-J, Lee S, Shim H, Chun Y, Choi W, et al. Flexible displays: Low-power flexible organic light-emitting diode display device. Advanced Materials. 2011;**23**(31):3475-3475. DOI: 10.1002/adma.201190120

[13] Brotzman, R, Paskiewicz D-M. U.S. Patent Application No.10/033,015; 2018

[14] Chen C-J, Lin K-L. Internal stress and adhesion of amorphous Ni–Cu–P alloy on aluminum. Thin Solid Films. 2000;**370**(1-2):106-113. DOI: 10.1016/S0040-6090(00)00859-2

[15] Li L, Tirrell M, Korba GA, Pocius AV. Surface energy and adhesion studies on acrylic pressure sensitive adhesive. The Journal of Adhesion. 2001;**76**:307-334. DOI: 10.1080/00218460108030724

[16] Fang C, Jing Y, Zong Y, Lin Z. Effect of trifunctional cross-linker triallyl

isocyanurate (TAIC) on the surface morphology and viscoelasticity of the acrylic emulsion pressure-sensitive adhesives. Journal of Adhesion Science and Technology. 2017;**31**(8):858-873. DOI: 10.1080/01694243.2016.1235005

[17] Tse MF. Studies of triblock copolymer-tackifying resin interactions by viscoelasticity and adhesive performance. Journal of Adhesion Science and Technology. 1989;**3**(7):551-570. DOI: 10.1163/156856189X00407

[18] Tse MF, Jacob L. Pressure sensitive adhesives based on vector sis polymers I. rheological model and adhesive design pathways. The Journal of Adhesion. 1996;**56**(1-4):79-95. DOI: 10.1080/00218469608010500

[19] Gdalin BE, Bermesheva EV, Shandryuk GA, Feldstein MM. Effect of temperature on probe tack adhesion: Extension of the Dahlquist criterion of tack. The Journal of Adhesion. 2011;**87**(2):111-138. DOI: 10.1080/00218464.2011.545325

[20] Krenceski MA, Johnson JF, Temin SC. Chemical and physical factors affecting performance of pressure-sensitive adhesives. Journal of Macromolecular Science - Reviews in Macromolecular Chemistry and Physics. 1986;**C26**(1):143-182. DOI: 10.1080/07366578608081971

[21] Kowalski A, Czech Z, Byczynski L. How does the surface free energy influence the tack of acrylic pressure-sensitive adhesives (PSAs)? Journal of Coatings Technology and Research. 2013;**10**(6):879-885. DOI: 10.1007/s11998-013-9522-2

[22] Czech Z, Goracy K. Characterization of the crosslinking process of silicone pressure-sensitive adhesives. Polymer. 2005;**50**(10):762-764. DOI: 10.14314/polimery.2005.762

[23] Lin SB, Durfee LD, Ekeland RA, McVie J, Schalau GK. Recent advances in silicone pressure-sensitive adhesives. Journal of Adhesion Science and Technology. 2007;**21**(7):605-623. DOI: 10.1163/156856107781192274

[24] Park C-H, Lee S-J, Lee T-H, Kim H-J. Characterization of an acrylic polymer under hygrothermal aging as an optically clear adhesive for touch screen panels. International Journal of Adhesion and Adhesives. 2015;**63**:137-144. DOI: 10.1016/j.ijadhadh.2015.08.012

[25] Fang C, Jing Y, Zong Y, Lin Z. Effect of N,N-dimethylacrylamide (DMA) on the comprehensive properties of acrylic latex pressure sensitive adhesives. International Journal of Adhesion and Adhesives. 2016;**71**:105-111. DOI: 10.1016/j.ijadhadh.2016.09.003

[26] Kuo C-FJ, Chen J-B, Chang S-H. Low corrosion optically clear adhesives for conducting glass: I. effects of N,N-diethylacrylamide and acrylic acid mixtures on optically clear adhesives. Journal of Applied Polymer Science. 2018;**135**(21):46277. DOI: 10.1002/app.46277

[27] Czech Z, Milker R. Solvent-free radiation-curable polyacrylate pressure-sensitive adhesive systems. Journal of Applied Polymer Science. 2003;**87**(2):182-191. DOI: 10.1002/app.11303

[28] Tanase H, Mamiya J, Suzuki M. A new backlighting system using a polarizing light pipe. IBM Journal of Research and Development. 1998;**42**(3.4):527-536. DOI: 10.1147/rd.423.0527

[29] Park CH, Lee SJ, Lee TH, Kim HJ. Characterization of an acrylic pressure-sensitive adhesive blended with hydrophilic monomer exposed to hygrothermal aging: Assigning cloud point resistance as an optically clear adhesive for a touch screen panel. Reactive and Functional Polymers.

2016;**100**:130-141. DOI: 10.1016/j.reactfunctpolym.2016.01.012

[30] Baek SS, Hwang SH. Preparation of biomass-based transparent pressure sensitive adhesives for optically clear adhesive and their adhesion performance. European Polymer Journal. 2017;**92**:97-104. DOI: 10.1016/j.eurpolymj.2017.04.039

[31] Ishinabe T, Miyashita T, Uchida T. Wide-viewing-angle polarizer with a large wavelength range. Japanese Journal of Applied Physics. 2002;**41**:4553-4558. DOI: 10.1143/JJAP.41.4553

[32] Yeh P. Leakage of light in liquid crystal displays and birefringent thin film compensators. Optical Review. 2009;**16**(2):192-198. DOI: 10.1007/s10043-009-0035-2

[33] Cristaldi DJ, Pennisi S, Pulvirenti F. Liquid Crystal Display Drivers: Techniques and Circuits. New York: Springer; 2009

[34] Koden M. Wide viewing angle technologies of TFT-LCDs. Sharp Technical Journal. 1999;**74**(8):55-60

[35] Ma J, Ye X, Jin B. Structure and application of polarizer film for thin-film-transistor liquid crystal displays. Displays. 2011;**32**(2):49-57. DOI: 10.1016/j.displa.2010.12.006

[36] Kumar RS, Auch M, Ou E, Ewald G, Jin CS. Low moisture permeation measurement through polymer substrates for organic light emitting devices. Thin Solid Films. 2002;**417**(1-2):120-126. DOI: 10.1016/S0040-6090(02)00584-9

[37] Asnafi N. On springback of double-curved autobody panels. International Journal of Mechanical Sciences. 2001;**43**(1):5-37

[38] Campbell CJ, Clapper J, Behling RE, Erdogan B, Beagi HZ, Abrahamson JT, et al. Optically clear adhesives enabling foldable and flexible OLED displays. SID Symposium Digest of Technical Papers. 2017;**48**(1):2009-2011. DOI: 10.1002/sdtp.12071

[39] Kao SC, Li LJ, Hsieh MC, Zhang S, Tsai PM, Sun ZY, et al. 71-1: Invited paper: The challenges of flexible OLED display development. SID Symposium Digest of Technical Papers. 2017;**48**(1):1034-1037. DOI: 10.1002/sdtp.11812

[40] Yan J-Y, Ho J-C, Chen J. Foldable AMOLED display development: Progress and challenges. Information Display. 2018;**31**(1):12-16. DOI: 10.1002/j.2637-496X.2015.tb00780.x

[41] Nishimura M, Takebayashi K, Hishinuma M, Yamaguchi H, Murayama A. A 5.5-inch full HD foldable AMOLED display based on neutral-plane splitting concept. Journal of the Society for Information Display. 2019;**27**(8):480-486. DOI: 10.1002/jsid.796

[42] Su Y, Li S, Li R, Dagdeviren C. Splitting of neutral mechanical plane of conformal, multilayer piezoelectric mechanical energy harvester. Applied Physics Letters. 2015;**107**:041905. DOI: 10.1063/1.4927677

[43] SAMSUNG Newsroom. [Video] New Form Factor, New Rules: Watch the Galaxy Fold's Folding Test [Internet]. Available from: https://news.samsung.com/global/video-new-form-factor-new-rules-watch-the-galaxy-folds-folding-test [Accessed: 11 August 2019]

[44] Komatsu R, Nakazato R, Sasaki T, Suzuki A, Senda N, Kawata T, et al. Repeatedly foldable AMOLED display. Journal of the Society for Information Display. 2018;**23**(2):41-49. DOI: 10.1002/jsid.276

[45] Abrahamson J, Beagi H, Salmon F, Campbell C. Optically clear adhesives for OLED. In: OLED Technology and

Applications. London: IntechOpen; 2019. DOI: 10.5772/intechopen.88659

[46] Cheng A, Chen Y, Jin J. Su T. study on mechanical behavior and effect of adhesive layers in foldable AMOLED display by finite element analysis. SID Symposium Digest of Technical Papers. 2019;**50**(1):1060-1063. DOI: 10.1002/sdtp.13110

[47] Lin L, Dang P, Hu K, Gao X, Huang X. Challenges and Progress of small bending radius foldable AMOLED display module technology. SID Symposium Digest of Technical Papers. 2017;**48**(1):445-446. DOI: 10.1002/sdtp.11673

[48] 3M. Substrates and Adhesion [Internet]. Available from: https://www.3m.com/3M/en_US/bonding-and-assembly-us/resources/full-story/?storyid=1d2481ca-5c8c-455d-952d-5ed90e04e8a7 [Accessed: 11 August 2019]

[49] Kim HJ, Park YJ, Choi JH, Han HS, Hong YT. Surface modification of polyimide film by coupling reaction for copper metallization. Journal of Industrial and Engineering Chemistry. 2009;**15**(1):23-30. DOI: 10.1016/j.jiec.2008.08.016

[50] Bichler CH, Langowski HC, Moosheimer U, Seifert B. Adhesion mechanism of aluminum, aluminum oxide, and silicon oxide on biaxially oriented polypropylene (BOPP), poly (ethyleneterephthalate)(PET), and poly (vinyl chloride)(PVC). Journal of Adhesion Science and Technology. 1997;**11**(2):233-246. DOI: 10.1163/156856197X00336

[51] Sperling LH. Introduction to Physical Polymer Science. 4th ed. Vol. 508-509. New Jersey: John Wiley & Sons; 2006. p. 530

[52] Lee JH, Lee TH, Shim KS, Park JW, Kim HJ, Kim Y, et al. Molecular weight and crosslinking on the adhesion performance and flexibility of acrylic PSAs. Journal of Adhesion Science and Technology. 2016;**30**(21):2316-2328. DOI: 10.1080/01694243.2016.1182382

[53] Rubinstein M, Colby RH. Polymer Physics. New York: Oxford University Press Inc.; 2003. p. 290

[54] Lee J-H, Lee T-H, Shim K-S, Park J-W, Kim H-J, Kim Y, et al. Effect of crosslinking density on adhesion performance and flexibility properties of acrylic pressure sensitive adhesives for flexible display applications. International Journal of Adhesion and Adhesives. 2017;**74**:137-143. DOI: 10.1016/j.ijadhadh.2017.01.005

[55] Lee J-H, Lee T-H, Shim K-S, Park J-W, Kim H-J, Kim Y. Adhesion performance and recovery of platinum catalyzed silicone PSAs under various temperature conditions for flexible display applications. Materials Letters. 2017;**208**:86-88. DOI: 10.1016/j.matlet.2017.05.042

[56] Lee J-H, Shim G-S, Park J-W, Kim H-J, Kim Y. Adhesion performance and recovery of acrylic pressure-sensitive adhesives thermally crosslinked with styrene–isoprene–styrene elastomer blends for flexible display applications. Journal of Industrial and Engineering Chemistry. 2019;**78**:461-467. DOI: 10.1016/j.jiec.2019.05.019

[57] Salmon F, Everaerts A, Campbell C, Pennington B, Erdogan-Haug B, Gregg C. Modeling the mechanical performance of a foldable display panel bonded by 3M optically clear adhesives. SID Symposium Digest of Technical Papers. 2017;**48**(1):938-941. DOI: 10.1002/sdtp.11796

[58] Jia Y, Liu Z, Wu D, Chen J, Meng H. Mechanical simulation of foldable AMOLED panel with a module structure. Organic Electronics. 2019;**65**:185-192. DOI: 10.1016/j.orgel.2018.11.026

Chapter 6

Numerical Simulation and Compact Modeling of Thin Film Transistors for Future Flexible Electronics

Arun Dev Dhar Dwivedi, Sushil Kumar Jain, Rajeev Dhar Dwivedi and Shubham Dadhich

Abstract

In this chapter, we present a finite element method (FEM)-based numerical device simulation of low-voltage DNTT-based organic thin film transistor (OTFT) by considering field-dependent mobility model and double-peak Gaussian density of states model. Device simulation model is able to reproduce output characteristics in linear and saturation region and transfer characteristics below and above threshold region. We also demonstrate an approach for compact modeling and compact model parameter extraction of organic thin film transistors (OTFTs) using universal organic TFT (UOTFT) model by comparing the compact modeling results with the experimental results. Results obtained from technology computer-aided design (TCAD) simulation and compact modeling are compared and contrasted with experimental results. Further we present simulations of voltage transfer characteristic (VTC) plot of polymer P-channel thin film transistor (PTFT)-based inverter to assess the compact model against simple logic circuit simulation using SmartSpice and Gateway.

Keywords: OTFTs, numerical simulation, compact modeling, flexible electronics

1. Introduction

The interest in organic thin film transistors (OTFTs) has increased significantly in the past few years and has been proven for various applications such as flexible low-cost displays [1], organic memory [2], key components of RFID [3] tags, low-end electronic products, and polymer circuits and sensors [4]. Flexible electronics is a new technology that builds electronic circuits by depositing electronic products on flexible substrates like plastics, paper, and even cloth. Compared with inorganic electronics, organic or flexible electronics have the various following advantages. First, it can be manufactured at a very low cost at low temperatures. Second, it is thin, lightweight, foldable, and bendable and has a strong light absorption, no crushing, mechanical flexibility, low energy consumption, and high emission efficiency. Third, the cost is lower due to cheaper materials and lower-cost deposition processes [5], and it is also used for large area applications. Actually a stack of organic semiconductors (OSC) and low-temperature polymer gate dielectrics and the rapid annealing process are suitable with high-throughput, low-cost printing manufacturing [6]. Researchers replaced semiconductors with organic materials such as DNTT [7], poly(3-octylthiophene) (P3OT), poly(3-hexylthiophene) (P3HT), and poly(3-alkylthiophene) layers, and dielectric layers are used to create complete flexibility. A bigger challenge is to enhance the real performance of organic devices so as to expand their usage in real-time commercial applications [8]. To enhance the speed of the device, a very great deal of the research efforts has been dedicated to increasing the mobility of organic materials by improving the deposition conditions [9]. In addition to mobility, other methods of improving OTFT performance include scaling the length of channel and changing the active layer thickness. The OTFT is usually fabricated in an inverted structure with gate at the bottom, and source and drain will be at the top. Gundlach et al. [10] show that the bottom contact structure has a strong dependence on the contact barrier and due to the different nature of the interface between the channel and the insulator, the device exhibits different electrical properties [11]. Recently, for p-type OSCs, very high mobility values of several tens of $cm^2\ V^{-1}\ s^{-1}$ have been reported for polymers and small molecules indicating that OSC has great potential for improved performance through chemical structures and process optimization [12]. In addition to performance, deep understanding of instability issues of OTFTs and finding stable and reliable solutions for OTFT is therefore very important [13]. Since the systematic cost of experimental investigation is very high and it requires a lot of time also, technology computer-aided design (TCAD) simulation of semiconductor devices is becoming very important for investigating the design and electrical characteristics of the device prior to fabrication of the device. Organic semiconductor technology has emerged in the past 20 years. Depending on the model, these devices have been developed and studied over the past decade. Compared to the silicon industry, for which public model is clearly defined and commonly used to provide designers with a relatively good description of the process, organic transistors still lack to have complete device models that can fully describe their electrical characteristics. Therefore TCAD simulation and compact modeling of organic transistors become very important.

In this chapter we present 2D device simulation of low-voltage DNTT-based OTFT using Silvaco's ATLAS 2D simulator which uses Poisson semiconductor device equation [14–22] continuity equation for charge carriers, drift diffusion transport model, and density of defect states model for simulation electrical characteristics of the device. Silvaco's UTMOST IV model parameter extraction software is used to get compact model parameters using UOTFT model [23]. Also TCAD simulation results and compact modeling results were compared and contrasted with the experimentally measured results of the device. Compact model has been applied for logic circuit simulation, and voltage transfer characteristics of PTFT-based inverter circuit have been simulated using the compact model parameters extracted from UOTFT model. This chapter contains five sections. This section introduces the content of the paper. The device structure and simulation are described in Section 2. The compact modeling, model verification, and parameter extraction are explained in Section 3. The results and discussion are explained in Section 4. Finally, conclusions drawn are given in Section 5.

2. Simulation

2.1 Device structure and finite element-based numerical simulation

The OTFT is designed on the bottom-gate top-contact of a flexible PEN substrate. A gate dielectric composed of a 3.6-nm-thick aluminum oxide layer and a

1.7-nm-thick n-tetradecylphosphonic acid self-assembled monolayer (SAM) was used [24]. Next, an organic semiconductor layer having a thickness of 11 nm was placed on the AlOx/SAM gate dielectric. The AlOx/SAM gate dielectric (5.3 nm) is very small in thickness and has a large capacitance per unit area, so transistors and circuits can operate at a low voltage of about 3 V. The OTFT has a channel length of 200 μm and a channel width of 400 μm, Lov = 10 μm.

The energy band diagram of a metal insulator semiconductor (MIS) structure is given in **Figure 1**. Maximum valence band energy (E_V) and minima of conduction band energy (E_C) of the inorganic semiconductor are substantially similar between the HOMO and the LUMO of the organic semiconductor. Especially for DNTT,

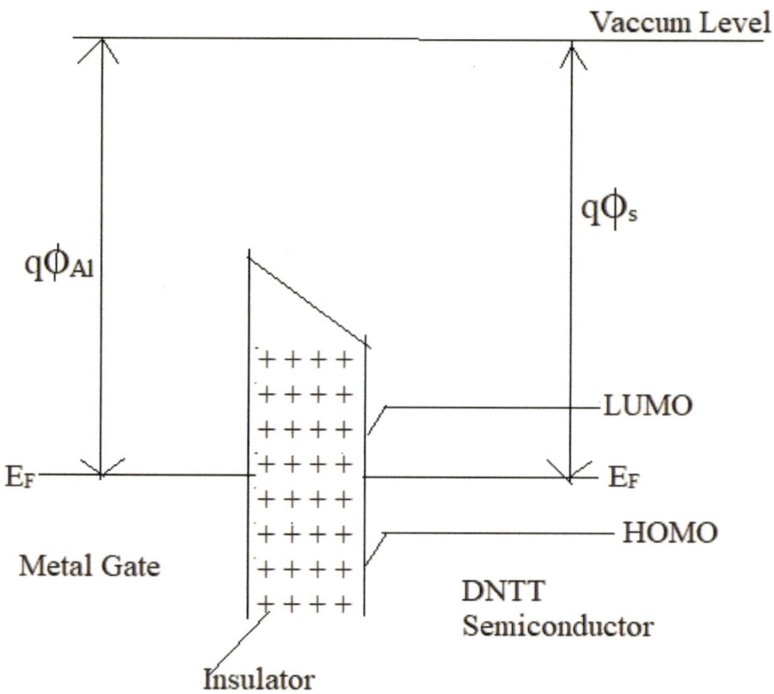

Figure 1.
Energy band diagram of a metal insulator semiconductor (MIS) structure.

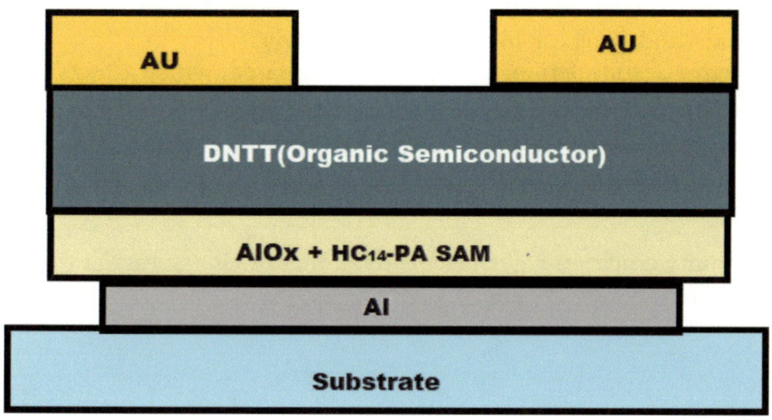

Figure 2.
Schematic cross-sectional diagram of organic TFTs along with the chemical structure of SAM and organic semiconductor.

HOMO is approximately −5.19 eV, and LUMO is about −1.81 eV [7, 24]. This introduces a large enough 3.38 eV HOMO-LUMO energy gap, which is sufficient for transistor operation.

To start the ATLAS simulation, we defined the physical structure and device dimensions, including the location of the electrical contacts. **Figure 2** shows a cross-sectional view of the bottom-gate, top-contact DNTT-based OTFT.

2.2 Device physical equation

We can calculate these charge carrier densities by solving basic device equations simultaneously including Poisson equation [14–22], electron and hole continuity equation, charge transfer equation, and defect density of states equation. The first three equations are the default equations that ATLAS uses to find the electrical behavior of the device.

The Poisson equation determines the electric field intensity in the given device based on the internal movement of the carriers and the distribution of the fixed charges given by Eq. (1):

$$\nabla . E = -\mathrm{div}(\nabla \Psi) = \frac{\rho(x, y)}{\epsilon} \tag{1}$$

where ϵ is the permittivity of the region and $\rho(x,y)$ is the charge density given by.

$$\rho(x, y) = q[p(x, y) - n(x, y) + N_D^+ (x, y) - N_A^- (x, y)] \tag{2}$$

where p(x,y) is the hole density, n(x,y) is the electron density, $N_D^+(x,y)$ is the ionization donor density, and N_A^- (x,y) is the ionization acceptor density.

To account for the trapped charge, Poisson equations are modified by adding an additional term Q_T, representing trapped charge given in (A):

$$\mathrm{div}(\varepsilon \nabla \Psi) = -\rho(x, y) = q[n(x, y) - p(x, y) - N_D^+ (x, y) + N_A^- (x, y)] - Q_T \tag{3}$$

where $Q_T = q(N^+_{tD} + N^-_{tA})$, N^+_{tD} = density × F_{tD}, and N^-_{tA} = density × F_{tA}. Here, N^+_{tD} and N^-_{tA} are ionized density of donor-like trap and ionized density of acceptor-like traps, respectively, and F_{tD} and F_{tA} are probability of ionization of donor-like traps and acceptor-like traps, respectively.

Due to charge accumulation, a potential is generated, which affects the intensity of electric field distribution and current. The voltage applied to the gate electrode generates an electric field that attracts a few or majority carriers. In addition, for OTFTs, the voltage potential between the source and the drain establishes another electric field along the channel that drives the charge carriers and produces a current.

The continuity equation describes the dynamics of charge carrier distribution over time as given in Eqs. (4) and (5):

$$\frac{\partial n}{\partial t} = \frac{1}{q} \nabla . J_n + G_n - R_n \tag{4}$$

$$\frac{\partial p}{\partial t} = -\frac{1}{q} \nabla . J_p + G_p - R_p \tag{5}$$

In these equations, q is the magnitude of the electronic charge, n is the electron carrier density, p is the hole carrier density, J is the corresponding current density, G is the corresponding charge generation rate, and R is the corresponding charge recombination rate. For organic/metal oxide semiconductor field-effect transistors (MOSFETs), there is no optical absorption, so the term is simplified and the properties of the material are described by the minority carrier recombination lifetime. Since MOSFETs are majority carrier devices, the characteristics of carrier generation and recombination are relatively unimportant. The physical properties of organic semiconductors depend on the generation and movement of polarons [25].

A third important set of equations for describing the device physics for the charge carrier is given by

$$J_p = \mathrm{qn}\mu_p E - qD_p \nabla p \tag{6}$$

$$J_n = \mathrm{qn}\mu_n E + qD_n \nabla n \tag{7}$$

It contains drift and diffusion parts. These equations determine the current density based on the carrier mobility (μ), the electric field (E), the carrier density (n, p), and the diffusion coefficient of the carrier (D). Diffusion coefficient operators are related to Einstein's mobility relationship:

$$D_n = \frac{kT}{q}\mu_n \tag{8}$$

$$D_p = \frac{kT}{q}\mu_p \tag{9}$$

In summary, the ATLAS software solves Poisson equations, continuity equations, and current density equations [26, 27] at each node in a two-dimensional grid for a given device structure simultaneously with itself and is subject to boundary conditions (including those applied at the contacts). With the help of ATLAS, the electric field distribution and electron and hole current density are calculated at each node and terminal current at electrode.

2.3 Density of defect states model

The assumed total density (DOS), g(E), consists of four bands: two tail bands (analogous to acceptor-like conduction band and donor-like valence band) and two deep energy bands (one donor-like and the other acceptor-like); they are modeled using a Gaussian distribution [15–22, 28]:

$$g(E) = g_{TA}(E) + g_{TD}(E) + g_{GA}(E) + g_{GD}(E) \tag{10}$$

Here, E is the trap energy, *EC* is the conduction band energy, *EV* is the valence band energy, and the subscripts (*T*, *G*, *A*, *D*) stand for tail, Gaussian (deep level), acceptor, and donor states, respectively:

$$g_{TA}(E) = NTA \exp\left[\frac{E - E_c}{WTA}\right] \tag{11}$$

$$g_{TD}(E) = NTD \exp\left[\frac{E_v - E}{WTD}\right] \tag{12}$$

$$g_{GA}(E) = NGA \exp\left[-\left[\frac{EGA - E}{WGA}\right]^2\right] \tag{13}$$

$$g_{GD}(E) = NGD \exp\left[-\left[\frac{E - EGD}{WGD}\right]^2\right] \tag{14}$$

For exponential tails, DOS is defined by its conduction and valence band edge intercept densities (NTA and NTD) and by its characteristic attenuation energy (WTD and WTA).

For Gaussian distribution, DOS is given by the total state density (NGD and NGA), its characteristic attenuation energy (WGD and WGA), and its peak energy distribution (EGD and EGA).

2.4 Trapped carrier density

The ionized densities of donor and acceptor states are given by Eqs. (14) and (15):

$$n_T = n_{TD} + n_{GD} \tag{15}$$

$$p_T = p_{TA} + p_{GA} \tag{16}$$

where p_{TA}, p_{GA}, n_{TD}, and n_{GD} are given below

$$p_{TA} = \int_{Ev}^{Ec} g_{TA}(E).f_{t_{TA}}(E, n, p)dE \tag{17}$$

$$p_{GA} = \int_{Ev}^{Ec} g_{GA}(E).f_{t_{GA}}(E, n, p)dE \tag{18}$$

$$n_{TD} = \int_{Ev}^{Ec} g_{TD}(E).f_{tTD}(E, n, p)dE \tag{19}$$

$$n_{GD} = \int_{Ev}^{Ec} g_{GD}(E).f_{tGD}(E, n, p)dE \tag{20}$$

$f_{tGA(E,n,p)}$ and $f_{tTA(E,n,p)}$ are the ionization probabilities for the Gaussian acceptor and tail DOS, while $f_{tTD(E,n,p)}$ and $f_{tGD(E,n,p)}$ are defined as the probability of occupation of a trap level at energy E for the Gaussian and tail acceptor, and donor states in steady state are given by following equations [24–27]:

$$f_{tTA}(E, n, p) = \frac{v_n SIGTAE.n + v_p SIGTAH.n_i \exp\left[\frac{E_i - E}{kT}\right]}{v_n SIGTAE\left(n + n_i \exp\left[\frac{E_i - E}{kT}\right]\right) + v_p SIGTAH\left(p + n_i \exp\left[\frac{E_i - E}{kT}\right]\right)} \tag{21}$$

$$f_{tGA}(E, n, p) = \frac{v_n SIGGAE.n + v_p SIGGAH.n_i \exp\left[\frac{E_i - E}{kT}\right]}{v_n SIGGAE\left(n + n_i \exp\left[\frac{E_i - E}{kT}\right]\right) + v_p SIGGAH\left(p + n_i \exp\left[\frac{E_i - E}{kT}\right]\right)} \tag{22}$$

$$f_{tTD}(E,n,p) = \frac{v_p\ SIGTDH.p + v_n\ SIGTDE.n_i \exp\left[\frac{E-E_i}{kT}\right]}{v_n SIGTDE\left(n + n_i \exp\left[\frac{E-E_i}{kT}\right]\right) + v_p SIGTDH\left(p + n_i \exp\left[\frac{E_i-E}{kT}\right]\right)} \tag{23}$$

$$f_{tGD}(E,n,p) = \frac{v_p\ SIGGDH.p + v_n\ SIGGDE.n_i \exp\left[\frac{E-E_i}{kT}\right]}{v_n SIGGDE\left(n + n_i \exp\left[\frac{E-E_i}{kT}\right]\right) + v_p SIGGDH\left(p + n_i \exp\left[\frac{E_i-E}{kT}\right]\right)} \tag{24}$$

where v_n is the thermal velocity of electron, v_p is the thermal velocity of hole, and n_i is intrinsic carrier concentration. SIGGAE and SIGTAE are the electron capture cross sections subject to the Gaussian states and main tail, respectively. SIGGAH and SIGTAH are hole trap cross sections of the Gaussian states and acceptor tail, respectively. SIGTDE, SIGGDE, SIGTDH, and SIGGDH are the equivalent for donor states [8].

2.5 Poole-Frenkel mobility model

Firstly, Miller et al. [29] described the rate of monophonic jumps for simulating hopping in inorganic semiconductors. Later, Vissenberg et al. [30] studied the dependency related to carrier transport on the energy distribution and the jump distance in amorphous transistors, which further helps to find the carrier mobility. The very popular Poole-Frenkel mobility model [31] is given by

$$\mu_{n_{PF}(E)} = \mu_{n0} \exp\left(-\frac{DELTAEN.PFMOB}{kT_{neff}} + \left(\frac{\text{BETAN} \cdot \text{PFMOB}}{kT_{neff}} - \text{GAMMAN.PFMOB}\right)\sqrt{|E|}\right) \tag{25}$$

$$\mu_{p_{PF}(E)} = \mu_{p0} \exp\left(-\frac{DELTAEP.PFMOB}{kT_{peff}} + \left(\frac{\text{BETAP} \cdot \text{PFMOB}}{kT_{peff}} - \text{GAMMAP.PFMOB}\right)\sqrt{|E|}\right) \tag{26}$$

where $\mu_{p_{PF}(E)}$ and $\mu_{n_{PF}(E)}$ are the Poole-Frenkel mobilities for holes and electrons, respectively; μ_{n0} and μ_{p0} are defined as the zero-field mobilities for electrons and holes, respectively; and E is the electric field. DELTAEN.PFMOB and DELTAEP.PFMOB are the activation energy at zero electric field for electrons and holes, respectively. BETAN.PFMOB is the electron Poole-Frenkel factor, and BETAP.PFMOB is the hole Poole-Frenkel factor. *Tneff* is the effective temperature for electrons, and *Tpeff* is the effective temperature for holes.

3. Compact modeling, model parameter extraction, and model verification

The technology and operation of organic thin film transistors (OTFTs) have various unique features that require a dedicated compact TFT model. The important features of OTFT include operation in carrier accumulation mode, exponential density of states, interface traps and space charge-limited carrier transport, nonlinear parasitic resistance, source and drain contacts without junction isolation, dependence of mobility on career concentration, electric field, and temperature. The universal organic TFT (UOTFT) model [23] is a modeling expression that extends the uniform charge control model (UCCM) previously used for a-Si and poly-Si TFTs [23, 32] to OTFTs and introduces a general expression of modeling for conductivity of channel of OTFTs [30, 33]. In this way, the UOTFT model is applicable to various OTFT device architectures, specifications of material, and manufacturing technologies.

3.1 Model features

UOTFT model depends on a general-purpose compact modeling approach [23, 32], which provides smooth interpolation of drain currents between linear and saturated operating regions including channel length modulation effects and also provides the unified expression of the gate-induced charge in the conductive channel which is valid in all operating states. This model also gives a unified charge-based mobility description, drain-source current, and gate-to-source and gate-to-drain capacitances.

3.2 Model description

The control equation for the UOTFT model for the n-channel OTFT case is described here. The p-channel condition can be obtained by direct change in voltage, charge polarity, and current.

The charge accumulation in channel per unit area at zero-channel potential $(-Q_{acc})_o$ is calculated by the help of the solution of the UCCM equation given by [34].

$$(-Q_{acc})_o = C_i.V_{gse} \tag{27}$$

$$V_{gse} = \mathrm{VO(T)}.\ln\left[1+\frac{e^{u+1}}{1+k(u+2)\ln\left(1+e^{u+1}\right)}\right] \tag{28}$$

$$u = \frac{V_{gs} - VT(T)}{VO(T)} \tag{29}$$

$$k(x) = 1 - \frac{84.4839}{x^2 + 150.864} \tag{30}$$

$$C_i = \in_0 \frac{EPSI}{TINS} \tag{31}$$

where C_i is the gate insulator capacitance per unit area, Vgse is the effective intrinsic gate-source voltage, Vgs is the gate-source voltage (intrinsic), VT is the temperature-dependent threshold voltage parameter, and VO is characteristic voltage (temperature-dependent); for carrier density of states including the influence of interface traps, $\in_0$ is the vacuum permittivity, and EPSI and TINS are model parameters representing the relative permittivity and thickness of the gate insulator, respectively.

3.3 Effective channel mobility

For accurate modeling of OTFTs, we should consider the characteristic power-law dependence of mobility on carrier concentration. According to the results of percolation theory [30], effective channel mobility is expressed in the UOTFT model as

$$\mu_C = MUACC(T).\left(\frac{(-Q_{acc})_0}{C_i.VACC}\right)^{GAMMA(T)} \tag{32}$$

MUACC, VACC, and GAMMA are model parameters. MUACC is a temperature-related parameter which defines effective channel mobility at the onset of strong accumulation of channel. This onset point is controlled by model parameter VACC. The power-law dependence of the mobility on carrier concentration is defined by the temperature-dependent model parameter GAMMA.

3.4 Intrinsic drain-source current

Drain-source current of intrinsic transistor due to charge carriers accumulated in the channel is defined by general interpolation expressions [23]:

$$I_{ds}^{acc} = G_{ch}.V_{dse} \tag{33}$$

$$Vdse = \frac{V_{ds}}{\left[1 + \left(\frac{G.V_{ds}}{I_{sat}(1+LAMBDA.Vds)}\right)^{MSAT}\right]^{\frac{1}{MSAT}}} \tag{34}$$

where G_{ch} is the effective channel conductance in the linear region, V_{dse} is the effective intrinsic drain-source voltage, V_{ds} is the intrinsic drain-source voltage, parameter LAMBDA defines the finite output conductance in the saturation region, and MSAT is the model parameter that provides a smooth transition between linear and saturated transistor operation. I_{sat} is the ideal intrinsic drain-source saturation current, and the effective channel conductance in the linear region G_{ch} is obtained in the following way:

$$G_{Ch} = \frac{G_{ch0}}{1 + G_{cho}.R_{ds}} \tag{35}$$

$$G_{ch0} = \frac{W_{eff}}{L_{eff}}.\mu_{c.}(-Q_{acc})_0 \tag{36}$$

$$R_{ds} = \frac{RDS(T)}{1 + \frac{V_{gse}}{VRDS}} \tag{37}$$

where G_{ch0} is the intrinsic effective conductance of channel in the linear region and R_{ds} is the nonlinear bias-dependent series resistance for intrinsic channel region defined by temperature-dependent model parameter RDS and the model parameter VRDS; on the other hand, Weff and Leff are effective channel widths and length, respectively.

The drain saturation current I_{sat} is determined by the following formula:

$$I_{sat} = G_{ch}.V_{sat} \tag{38}$$

where V_{sat} is the saturation voltage obtained as

$$V_{sat} = \frac{ASAT(T)}{C_i}\left[\frac{(-Q_{acc})_0}{GAMMA(T) + 2} + \frac{C_i VO(T)}{GAMMA(T) + 1}\right] \tag{39}$$

where $ASAT$ is the temperature-dependent model parameter.

The drain-source leakage current is obtained as

$$I_{ds}^{leak} = \frac{w_{eff}}{L_{eff}}\{\mathrm{IOL(T)}.\left[\exp\left(\mathrm{NSDL}.\frac{V_{ds}}{V_{th}}\right) - 1\right].\exp\left(-\mathrm{NSGL}\frac{V_{gs}}{V_{th}}\right) + \mathrm{SIGMAO}.V_{ds} \tag{40}$$

The IOL is a temperature-dependent leakage saturation current model parameter; NGSL and NDSL are non-ideal factors for gate and drain bias, respectively, and SIGMAO is a model parameter representing zero-bias drain-source conductivity:

$$V_{th=}\frac{kT}{q} \tag{41}$$

where V_{th} is the thermal voltage at device operating temperature.

The total intrinsic drain-source current is

$$I_{ds} = I_{ds}^{acc} + I_{ds}^{leak} \tag{42}$$

4. Results and discussion

4.1 Material parameters used for DNTT

The DNTT-based OTFT is designed in a bottom-gate, top-contact configuration. The designed structure has a channel length of 200 μm and a channel width of 400 μm with L_{ov} = 10 μm as shown in **Figure 2**. For the simulation of DNTT-based OTFT structure, the following parameters [24] used are listed in **Table 1**.

4.2 Comparison of TCAD-based numerical simulation characteristics and compact model-based simulation characteristics with experimental characteristics

Figure 3 shows the transfer characteristics obtained for the TCAD-based numerical simulation, compact model-based simulation of DNTT-based organic thin film transistor, and the measured characteristic of DNTT-based OTFT [24]. The transfer characteristics are obtained by varying the gate-to-source voltage (V_{GS}) from 0 to −3 V keeping drain voltage constant at −2 V. There is very good agreement between TCAD-based numerical simulation, compact model-based simulation of the transfer characteristics of OTFT, and experimental transfer characteristics of the fabricated device. **Figure 4** shows the output characteristics obtained from the TCAD-based numerical simulation, compact model-based simulation of DNTT-based organic thin film transistor, and the measured output characteristics of DNTT-based OTFT [24]. Output characteristics are obtained by varying the

Material simulation parameters	Value
DNTT energy band gap (eV)	3.38 eV
Occupied molecular orbital Of DNTT (highest)	−5.19 eV
Occupied molecular orbital Of DNTT (lowest)	−1.81 eV
Intrinsic p-type doping in DNTT	10^{16} cm^{-3}
Fixed interface charge concentration	5×10^{16} cm^{-3}
Work function of aluminum gate	4.1 eV
Work function of Au contact	5.0 eV
Semiconductor thickness of DNTT	11 nm
Dielectric thickness	5.3 nm

Table 1.
Simulation parameters of material of the OTFT [17].

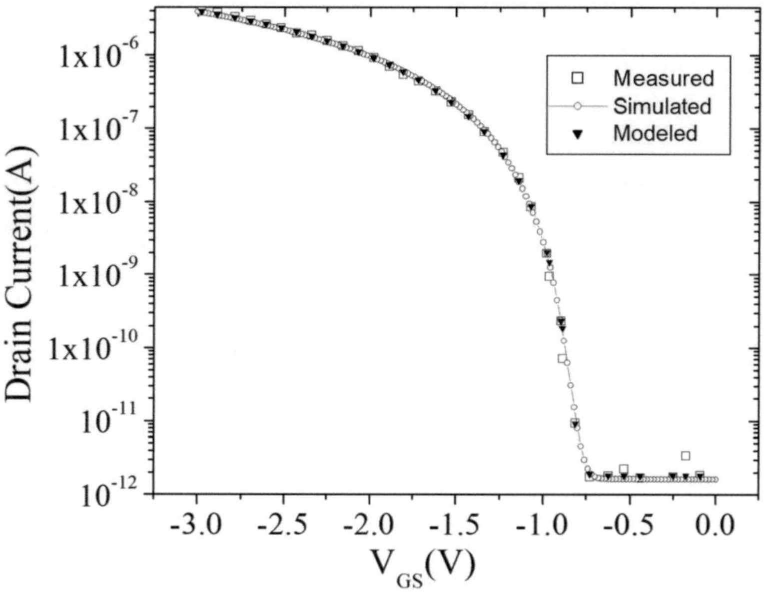

Figure 3.
Comparisons of transfer characteristics of the measured data, the TCAD-simulated data, and the modeled data.

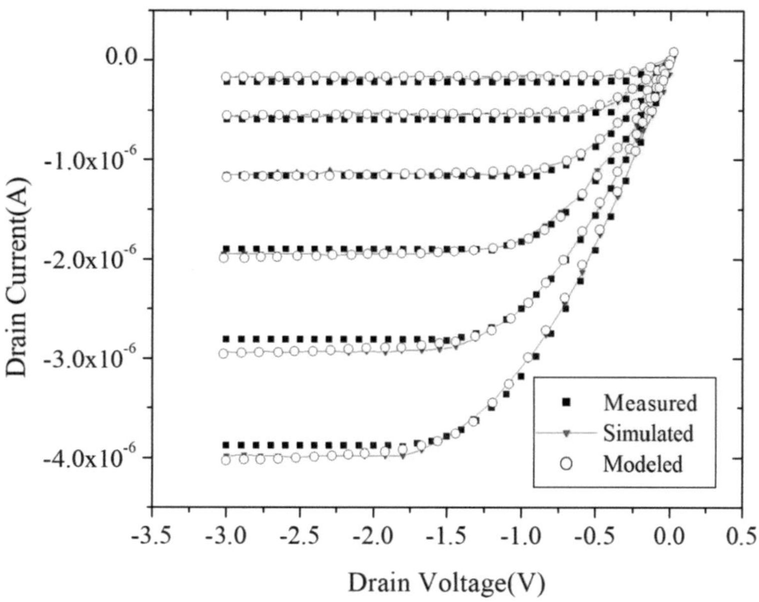

Figure 4.
Comparisons of output characteristics of the measured data, the TCAD-simulated data, and the modeled data.

drain-to-source voltage (V_{DS}) from 0 to −3 V and keeping the gate-to-source voltage (V_{GS}) constant at −1.5, −1.8, −2.1, −2.4, −2.7, and −3.0 V. The simulated output characteristic matched with the experimental output characteristic of the fabricated device.

4.3 Parameter extraction

Extracted OTFT model parameters for low-voltage DNTT-based OTFT using UOTFT model is given in **Table 2**. The extraction process starts with the collection

Parameter name	Symbol	UNIT	Typical values
The thickness of gate insulator	TINS	m	5.3×10^{-9}
Relative dielectric permittivity of the insulator at gate	EPSI	—	3.37
Relative dielectric permittivity of the semiconductor	EPS	—	3.0
Zero-bias threshold voltage	VT	V	−0.884542
Trap density states characteristic voltage	VO	V	0.0314021
Characteristic effective accumulation channel mobility	MUACC	cm^2/Vs	1.85
Characteristic voltage of the effective mobility	VACC	V	1.0
Output conductance parameter	LAMBDA	1/V	0.0
Knee-shape parameter	MSAT	—	5.0
Saturation modulation parameter	ASAT	—	1.52
Leakage saturation current	IOL	A	1×10^{-10}
Contact resistance	RS + RD	Kilo Ohm	116.892

Table 2.
Model parameters extracted for UOTFT model.

of data for I_D-V_G and I_D-V_D characteristic and providing it in UTMOST IV database in .uds format. Further we performed simulation of I_D-V_D and I_D-V_G characteristic using UOTFT model and optimization of this characteristic using Levenberg–Marquardt optimization technique with respect to experimental data for extraction of model parameters.

4.4 Simulation of logic circuit

For UOTFT model validity, simple logic circuit has been implemented and simulated based on p-type OTFTs only. The schematic in **Figure 5** shows the simple inverter circuit used in the simulation of a load transistor with auxiliary gate voltage

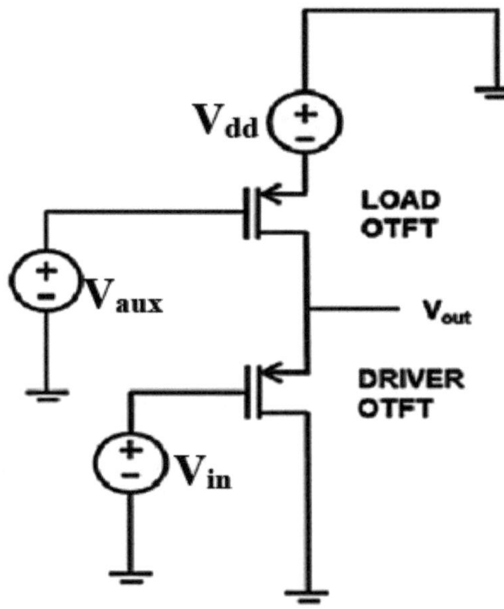

Figure 5.
A circuit diagram of the inverter circuit used for assessing the simulation results.

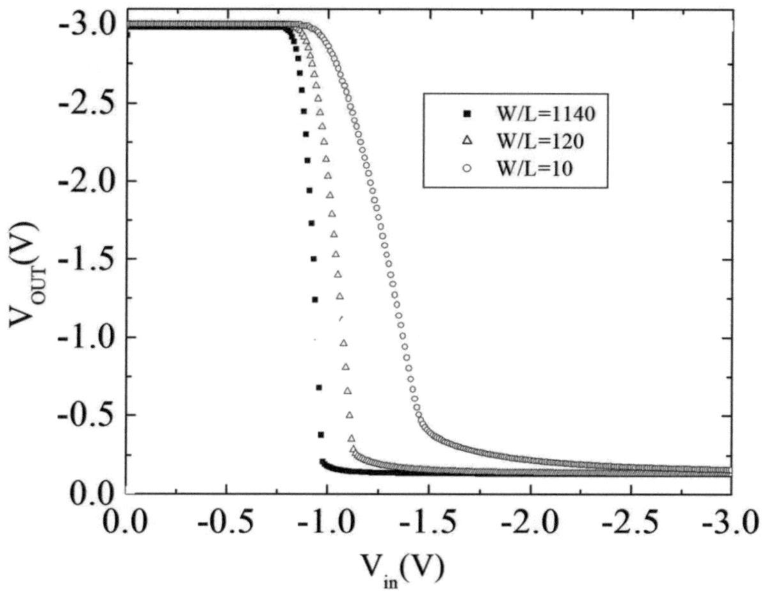

Figure 6.
Voltage transfer characteristics of inverter circuit shown for different W/L ratios of driver OTFT.

(V). The given inverter circuit works like a potential divider between the driver and the load OTFT. When the input voltage is lower than the threshold voltage (more positive than VT), the driver OTFT turns off. On the other side, when it is more than the threshold voltage (more negative than VT), the driver OTFT turns on. The operation of the inverter also depends on load TFT size relatively with the driver TFT. To assess whether the simulation correctly reproduces this dependence, the size of load OTFT and its gate voltage (V) remain at the same value, while the size and gate voltage of driver OTFT change. **Figure 6** shows the voltage transfer characteristic (VTC) plot of the inverter circuit under consideration for W/L ratio of 10, 120 1140 of driver TFT. As W/L ratio of the driver OTFT increases, its impedance decreases, and the transition between high and low states becomes clearer.

5. Conclusion

We presented a finite element method (FEM)-based device simulation of low-voltage DNTT-based OTFT by considering field-dependent mobility model and double-peak Gaussian density of states model using device simulator ATLAS. We also presented the application of UOTFT model and parameter extraction method to organic TFTs. We can also conclude that numerical simulations, experiments, and compact modeling-based simulation characteristics demonstrate the same behavior as matched in **Figure 3** and **Figure 4**. We simulated an OTFT based on DNTT and demonstrated the application of the UOTFT model to organic TFTs and also use experimental data from DNTT-based OTFTs to extract parameters for Silvaco's general-purpose organic TFT compact model. The model has been verified against logic circuit simulation. It has been concluded that UOTFTs provide more accurate modeling of the simpler parameter extraction methods for various organic TFTs. The results show that the UOTFT model correctly simulates the behavior of the devices reported in this study and is expected to be used for more complex circuits based on organic thin film transistors.

Acknowledgements

The authors are thankful to SERB, DST, Government of India, for the financial support under Early Career Research Award (ECRA) for Project No. ECR/2017/000179.

Author details

Arun Dev Dhar Dwivedi*, Sushil Kumar Jain, Rajeev Dhar Dwivedi and Shubham Dadhich
Department of Electrical and Electronics Engineering, Poornima University, Jaipur, India

*Address all correspondence to: adddwivedi@gmail.com

References

[1] Mizukami M, Hirohata N, Iseki T, Ohtawara K, Tada T, Yagyu S, et al. Flexible AMOLED panel driven by bottom-contact OTFTs. IEEE Electron Device Letters. 2006;**27**:249

[2] Takamiya M, Sekitani T, Kato Y, Kawaguchi H, Someya T, Sakurai T. An organic FET SRAM with back gate to increase static noise margin and its application to braille sheet display. IEEE Journal of Solid-State Circuits. 2007;**42**:93

[3] Cantatore E, Geuns TCT, Gelinck GH, Veenendaal EV, Gruijthuijsen AFA, Schrijnemakers L, et al. A 13.56 MHz RFID system based on organic transponers. IEEE Journal of Solid-State Circuits. 2007;**42**:84

[4] Brianda D, Opreab A, Courbata J, Barsanb N. Making environmental sensors on plastic foils. Materials Today. 2011;**14**:416

[5] Lodha A, Singh R. Prospects of manufacturing organic semiconductor-based integrated circuits. IEEE Transactions on Semiconductor Manufacturing. Aug 2001;**14**(3)

[6] Petti L et al. Metal oxide semiconductor thin-film transistors for flexible electronics. Applied Physics Reviews. 2016;**3**(2):021303

[7] Yamamoto T, Takimiya K. Facile synthesis of highly π-extended heteroarenes, dinaphtho[2,3-b,2′,3’f] chalcogenopheno[3,2-b] chalcogenophenes, and their application to field-effect transistors. Journal of the American Chemical Society. 2007; **129**(8):2224-2225

[8] Jeon J, Murmann B, Bao Z. Fully inkjet-printed short-channel organic thin-film transistors and inverter arrays on flexible substrates. IEEE Electron Device Letters. 2010;**31**:1488

[9] Han CY, Ma YX, Tang WM, Wang XL, Lai PT. A study on pentacene organic thin-film transistor with different gate materials on various substrates. IEEE Electron Device Letters. 2017;**38**(6)

[10] Gundlach DJ, Zhou L, Nichols JA, Jackson TN. An experimental study of contact effects in organic thin film transistors. Journal of Applied Physics. 2006;**100**:024509

[11] Street RA, Salleo A. Contact effects in polymer transistors. Applied Physics Letters. 2002;**81**:2887

[12] Lee BH, Bazan GC, Heeger AJ. Doping-induced carrier density modulation in polymer field-effect transistors. Advanced Materials. 2016; **28**(1):57-62

[13] Nikolka M et al. High operational and environmental stability of high-mobility conjugated polymer field-effect transistors through the use of molecular additives. Nature Materials. Dec. 2017;**16**:356-362. DOI: 10.1038/nmat4785

[14] Buonomo A, di Bello C. On solving Poisson's equation in two-dimensional semiconductor devices. Electronics Letters. 1984;**20**:4

[15] Dwivedi ADD. Numerical simulation and spice modeling of organic thin film transistors (OTFTs). International Journal of Advanced Applied Physics Research. 2014;**1**:14-21

[16] Vyas S, Dwivedi ADD, Dwivedi RD. Effect of gate dielectric on the performance of ZnO based thin film transistor. Superlattices and Microstructures. 2018;**120**:223-234

[17] Dwivedi ADD, Dwivedi RD, Dwivedi RD, Vyas S, Chakrabarti P. Numerical simulation of P3HT based

organic thin film transistors (OTFTs). International Journal of Microelectronics and Digital Integrated Circuits. 2015;**1**:13-20

[18] Kumari P, Dwivedi ADD. Modeling and simulation of pentacene based organic thin film transistors with organic gate dielectrics. Journal of Microelectronics and Solid State Devices. 2017;**4**:13-18

[19] Kushwah NS, Dwivedi ADD. Computer modeling of organic thin film transistors (OTFTs) using Verilog-A. Journal of Microelectronics and Solid State Devices. 2018;**5**:1-7

[20] Dwivedi ADD, Kumari P. Numerical simulation and characterization of pentacene based organic thin film transistors with top and bottom gate configurations. Global Journal of Research in Engineering-F. 2019;**19**:7-12

[21] Dwivedi ADD, Kumari P. TCAD simulation and performance analysis of single and dual gate OTFTs. Surface Review Letters. 2019;**1950145**:1-7. DOI: 10.1142/S0218625X19501452

[22] Dwivedi ADD, Dwivedi RD, Dwivedi RD, Zhao Q. Technology computer aided design (TCAD) based simulation and compact modeling of organic thin film transistors (OTFTs) for circuit simulation. International Journal of Advanced Applied Physics Research. 2019;**6**:1-5

[23] Fjeldly A, Ytterdal T, Shur M. Introduction to Device Modeling and Circuit Simulation. New York, NY, USA: Wiley; 1998

[24] Zaki T, Scheinert S, Hörselmann I, Rödel R, Letzkus F, Richter H, et al. Accurate capacitance modeling and characterization of organic thin-film transistors. IEEE Transactions on Electron Devices. 2014;**61**:98

[25] Hertel D, Bässler H. Photoconduction in amorphous organic solids. ChemPhysChem. 2008;**9**(5):666

[26] van Roosbroeck W. Theory of the flow of electrons and holes in germanium and other semiconductors. The Bell System Technical Journal. 1950; **29**:560

[27] Juengel A. Drift-Diffusion Equations. Springer; 2009. pp. 99-127. Chap. 5

[28] SILVACO. ATLAS User's Manual—Device Simulation Software. USA; 2010

[29] Miller A, Abrahams E. Impurity conduction at low concentrations. Physics Review. 1960;**120**:745

[30] Vissenberg MCJM, Matters M. Theory of the field-effect mobility in amorphous organic transistors. Physical Review B. 1998;**57**:964

[31] Shim CH, Maruoka F, Hattori R. Structural analysis on organic thin-film transistor with device simulation. IEEE Transactions on Electron Devices. 2010; **57**:195

[32] Iniguez B, Picos R, Veksler D, Koudymov A, Shur MS, Ytterdal T, et al. Universal compact model for long- and short-channel thin-film transistors. Solid-State Electronics. 2008;**52**:400

[33] Estrada M, Cerdeira A, Puigdollers J, Resendiz L, Pallares J, Marsal LF, et al. Accurate modeling and parameter extraction for organic TFTs. Solid-State Electronics. 2005;**49**:1009

[34] UTMOST IV Spice Models Manual. Santa Clara, CA, USA: Silvaco International; 2018

Chapter 7

Smart Manufacturing Technologies for Printed Electronics

Saleem Khan, Shawkat Ali and Amine Bermak

Abstract

Fabrication of electronic devices on different flexible substrates is an area of significant interest due to low cost, ease of fabrication, and manufacturing at ambient conditions over large areas. Over the time, a number of printing technologies have been developed to fabricate a wide range of electronic devices on nonconventional substrates according to the targeted applications. As an increasing interest of electronic industry in printed electronics, further expansion of printed technologies is expected in near future to meet the challenges of the field in terms of scalability, yield, and diversity and biocompatibility. This chapter presents a comprehensive review of various printing electronic technologies commonly used in the fabrication of electronic devices, circuits, and systems. The different printing techniques based on contact/noncontact approach of the printing tools with the target substrates have been explored. These techniques are assessed on the basis of ease of operation, printing resolutions, processability of materials, and ease of optimization of printed structures. The various technical challenges in printing techniques, their solutions with possible alternatives, and the potential research directions are highlighted. The latest developments in assembling various printing tools for enabling high speed and batch manufacturing through roll-to-roll systems are also explored.

Keywords: printed electronic technology, printed electronics, flexible electronics, large area electronics, roll to roll

1. Introduction

Printing electronics is a special type of manufacturing, where electronic components, circuits, and systems are developed on a wide variety of substrates in a similar fashion as drawing text and figures on a paper, textile, and handicrafts [1, 2]. The difference between normal printing and printed electronics is that in printed electronics, functional material is used as ink that exhibits functionalities of insulator, conductor, and semiconductor materials, which are essential for the electronic devices [3]. In most of the fabrication of electronic devices, these materials are sprayed over the substrate with the help of printing technology as low as few nanometers thick (thin film) and few micrometers width (pattern) [4, 5]. The combination of thin films and patterns can make any electronic device to be used in the electronic circuits [6]. Although the process seems simple, but the limitations posed by the various parameters such as uniformly dispersed and stable colloidal solutions, substrate treatments, and above all, optimized printing recipes make the printing process much more challenging [7, 8]. The prominent challenges in printed electronics are material compatibility, substrate surface energy, viscosity of the materials' solution, and compatibility of dissimilar materials in multilayer structure, technology limitations in terms of film thickness, width, and height. As compared to conventional electronic manufacturing, printing technologies are revolutionizing the incredible field of flexible/bendable electronics by providing cost-effective routes for processing diverse electronic materials on nonplanar substrates at compatible temperatures [9–12]. Simplified processing steps, reduced materials' wastage, low-fabrication costs, and simple patterning techniques make printing technologies very attractive when compared to standard microfabrication in clean room processes. Attractive features of the printed electronics have allowed researchers to explore new avenues for material processing to develop electronic devices, circuits, and systems on such surfaces, which are difficult to realize with the conventional wafer-based fabrication techniques [13, 14]. In accordance with the electronic industry roadmap, the research in this field is slowly inching toward a merge of well-established microelectronics and the age-old printing technologies. Traditionally, prepatterned parts of a printing module are brought in conformal contact to the target flexible (nonflexible) substrates, resulting in the transfer of functional inks/solutions on target surfaces [15–21]. The mechanisms to transfer inks/solutions on target substrates divide printing technologies into two main streams, where the surfaces come in physical contact in one, and in the second approach, ink is deposited through noncontacting the surfaces. The different printing technologies based on contact/noncontact approach are summarized in **Figure 1**. In contact printing process, the patterned structures with inked surfaces are interfaced physically at controlled pressures with the target substrate.

In noncontact process, the solution is dispensed through the openings or nozzles, and structures are defined by moving the substrate holder in a preprogrammed pattern. The contact-based printing technologies consist of gravure printing, gravure-offset printing, screen printing, flexographic printing, microcontact printing, nanoimprint, and dry transfer printing. The prominent noncontact printing techniques include slot-die coating, electrohydrodynamics, and inkjet printing. The noncontact printing techniques have received greater attractions due to their distinct capabilities such as simplicity, affordability, speed, adaptability to the fabrication process, reduced material wastage, high resolution of patterns, and easy control by adjusting few process parameters [6, 21–27].

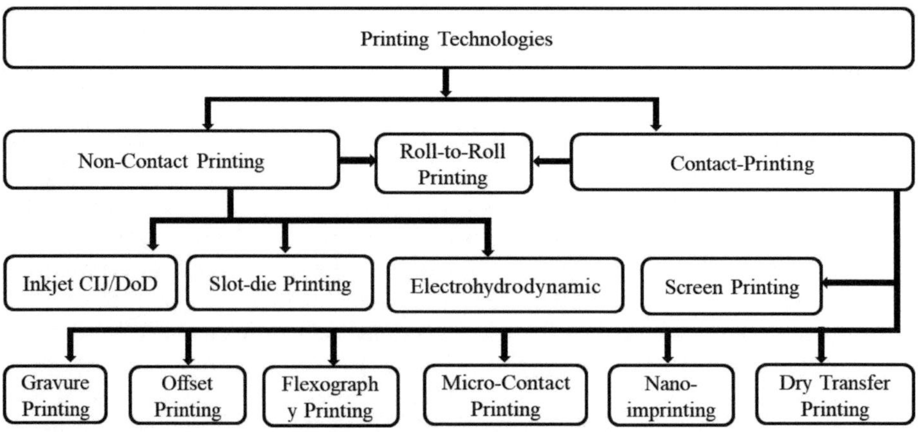

Figure 1.
The classification of common printing technologies.

2. Printed electronic technologies

The fascinating field of printed electronics is enabled by the rapid developments in printing technologies for precise deposition of functional inks in easy and cost-efficient way. The ultimate goal of developing the printed electronic technology is to revolutionize the device manufacturing and maximize the throughput by covering areas larger than wafer scales as well as increase the production through roll-to-roll processes. The attraction of printed electronic technology is that it can be executed at ambient conditions, thus enabling the fabrication of biocompatible electronics. With the help of printed electronic technologies and biocompatible materials, it is possible to fabricate electronic device on plastic substrates and even on human skin. The most commonly used and reliable printed electronic technologies are explained in the below sections.

2.1 Noncontact printing technologies

2.1.1 Inkjet material printer

Inkjet printing has gained a significant interest in recent years for processing solution-based nanomaterials and patterning on diverse substrates in a single step. Nanoparticles of the functional materials are mixed in compatible solvents to prepare printable ink. Besides, chemical-based solutions are also prepared and adjusted to the jetting parameters of the inkjet printing systems. Materials are ejected in micrometer-sized droplets through miniaturized nozzle printheads. Mainly two mechanisms for the actuation of inkjet nozzle head have been developed, i.e., thermal and piezoelectric. Droplets in very small diameters are ejected at each corresponding pulse and generated by either thermal or piezoelectric actuators used in the inkjet nozzle head. Schematic in **Figure 2** shows mechanism of droplet ejection through piezoelectric/thermal inkjet material printer. Typical nozzle diameters used for inkjet heads range from 10 to 150 μm and reservoir ink with a volume capacity of 3 ml.

The actuating element is either thermal or piezoelectric, in case of thermal actuator, a filament is embedded in the nozzle head, and upon the voltage application, a bubble is generated in the nozzle that pushes ink to the nozzle tip and generates droplets. In case of piezoelectric element, piezoelectric crystal is placed inside the nozzle, and upon the voltage application, piezoelectric crystal vibrates, and as a result, pressure exerts on the ink inside the nozzle and droplet generates. **Figure 2** shows the droplet actuation mechanism of inkjet material printer; ink is filled in the reservoir,

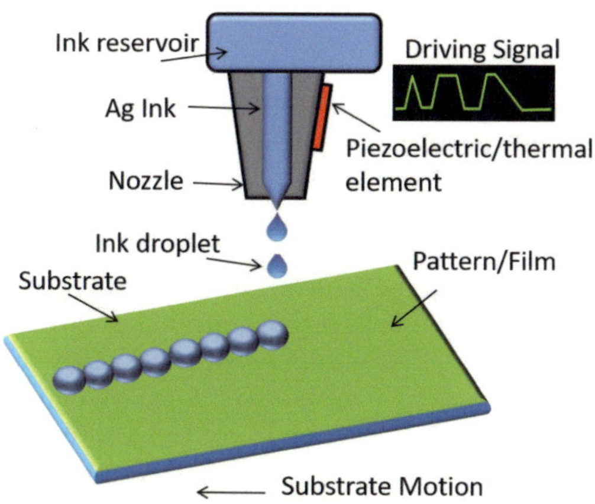

Figure 2.
Droplet's actuation mechanism of inkjet material printer.

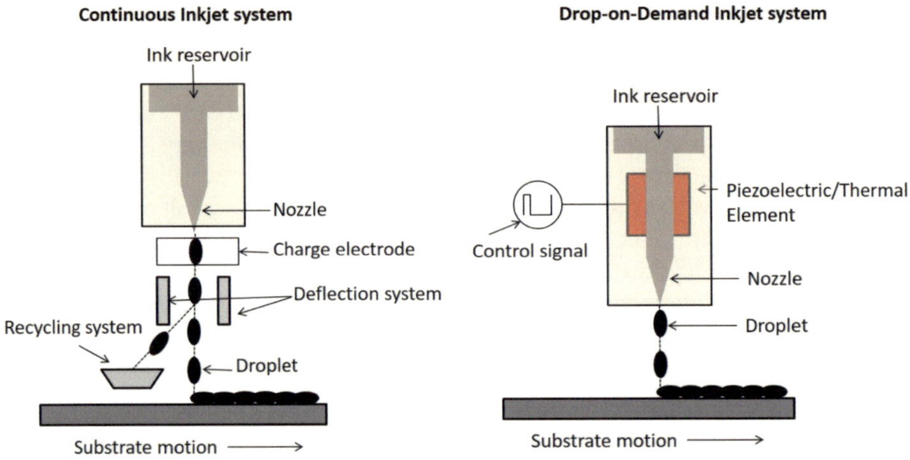

Figure 3.
Continuous and drop-on-demand inkjet systems.

and driving signal is applied to the actuating element that generates droplets according to the designed geometry. Inkjet printers are further divided into two types as continues inkjet (CIJ) and drop on demand (DoD) systems. In continuous inkjet printer, the ink is stored in a reservoir supplied to the nozzles. A charging electrode is used to pull ink out of the nozzle and form a droplet. After droplet formation, deflection plates are used to direct the droplets on the targeted area on the substrate. This kind of printing system is used in the industry for printing on the packages such as expiry date commonly seen on boxes and drinks of liquor bottles. As the droplets are not controlled by any input signal, the CIJ system continuously produces droplets with certain frequency; hence, the unused droplets are directed to the recycling system, where the ink is stored and reused. Schematic diagram of the CIJ system is shown in **Figure 3**. On the other hand, DoD system is controlled by a digital input signal that comes from the design of the pattern. In DoD system, each droplet is generated on demand; hence, there is no need of deflection and charge electrodes. The substrate is kept perpendicular to the nozzle, and droplets sit on the substrate one by one. The substrate is moved separately through a control system synchronized with the droplet generation in order to make a pattern or thin film on the substrate. The number of nozzles in commercial inkjet printhead ranges from 1 to 128 with ink droplets as small as 1.0 pL.

2.1.2 Electrohydrodynamic (EHD) printing

Electrohydrodynamics (EHD) is another interesting type of inkjet printing systems used to deposit functional ink in the form of thin films as well as high-resolution patterns. It is consisted of high-speed camera, light source, nozzle and head, ink storage and supply mechanism, stage movement, and activity display unit as shown in **Figure 4**. Working principle of the EHD system is that the ink to be deposited on substrate is pumped from ink storage tank to nozzle with appropriate ink flow rate to make stable cone jet [28]. Positive voltage is supplied to the nozzle, and ground is connected with substrate holder. The induction of the surface charges on the pendent meniscus emerging at the nozzle outlet results in an electric stress over the ink surface. If the electric field and flow rate are in some operating range, then this will overcome the surface tension stress over the ink surface and results in deformation of the droplet at the orifice of the nozzle into a conical shape. While sweeping the high voltage from zero to a required value, different modes of the spray occur including dripping, unstable, stable, and multi jet mode as shown in **Figure 4**. For stable cone

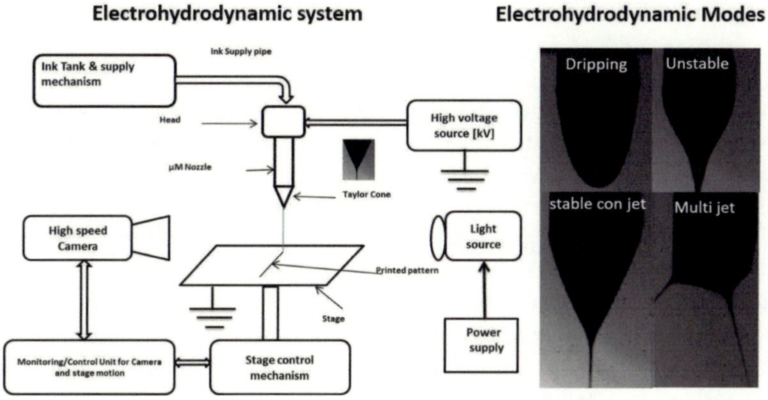

Figure 4.
Electrohydrodynamic system and different modes of jetting ink.

jet, appropriate voltage and flow rate are required, which result in uniform spray. The tangential electric field acting on the surface of the ink cone a thin jet emanates at the cone apex which further breaks up into a number of small droplets under the effect of coulomb forces [29–31]. In the stable cone jet mode, EHD is ideal for printing and can be used for thin films or pattern deposition on various substrates. The cone jet expands at the substrate sides and makes spray as the distance between nozzle and substrate, i.e., standoff is increased [32]. This distance needs to be optimized to achieve stable cone jet desired for thin film deposition [5]. Whereas, during small standoff distance, cone jet is narrow and used for patterning [11, 21].

2.1.3 Aerosol jet printing

Aerosol jet printing is an interesting technique in noncontact printing, which has attracted a significant interest in the manufacturing of high-resolution pattering. A wide variety of materials including insulators, semiconductors, and metallic conductors are processed in viscosity ranges of 1–1000 cps [1, 33]. The aerosol process is driven by the gas flows where a mist of microdroplets is generated as a result of pneumatic atomization or through ultrasonication. The capability to process a wide variety of materials and to pattern higher resolutions, as high as 10 μm on diverse substrates, makes aerosol jet printing most attractive among other contact-less printing techniques. Aerosol jet printing is divided into two main categories, i.e., pneumatic and ultrasonic, as shown in **Figure 5**. Both the techniques are based on different operational procedures and are used targeting specific set of requirements

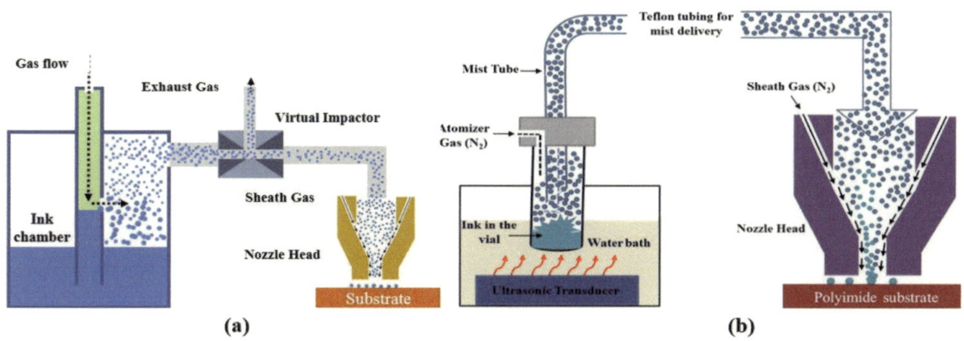

Figure 5.
Schematics of (a) pneumatic and (b) ultrasonic aerosol jet printing systems.

and goals [1, 34]. For instance, in pneumatic atomizer, aerosol mist is generated by supplying pressurized air/gas inside a closed chamber containing the ink, which results in the generation of microdroplets close at the ink and air interface. The microdroplets ranging in ≤5 μm diameter are entrained in the supplied gas and are driven toward the nozzle printhead. Whereas in the second case, the ink contained in a delicate vial is subjected to ultrasonication, and microdroplets are generated as a result of ultrasonic pressure waves. The microdroplets in the size ranges of ≤5 μm are entrained in the atomizer gas, which is driven toward the nozzle printhead. Another accompanying gas, i.e., sheath gas flow along with the nozzle orifice size, leads to define the pattern size to be printed on the target substrate. Pneumatic atomizer requires more materials, i.e., more than 10 mL, and is usually used for wide area printing. On the other hand, ultrasonic aerosol system requires about 0.5 mL of solution and can be used for printing at very high-resolution patterns, i.e., down to 10 μm, for instance. The annular gas flow reduces the size of the mist inside the carrier tube, which is accompanied by the sheath gas at the nozzle printhead. This sheath gas flow further converges the aerosol stream and reaches the surface at higher impact. The aerosol mist is ejected in the form of an intact jet, and the stage speed adjusted in such a way that the desired patterns are deposited in a continuous fashion.

2.1.4 Slot-die coater

Slot die is a special deposition technique, where the material is printed on a moving substrate and installed directly on a roll-to-roll system, as shown in **Figure 6**. The solution is directly dispensed off the slot-die printhead in a controlled manner. The slot-die coating is executed in two steps, where a uniform and stable flow of the coated material is achieved in the first phase. In the second phase, other processing parameters such as standoff distance between the slot-die opening and the target substrate speed of the rolling substrate and sintering conditions inline to promote a multilayer structure coating capability [35, 36]. System is connected to a continuous flow of materials into a temporary reservoir at the printhead and applied at appropriate amount on the rolling substrate as shown in **Figure 6**. This type of coating technique is ideal for large area deposition, such as solar cells and light emitting diodes, where high-resolution patterning is not required. Besides optimizing the rheological properties of the ink/solution, system specific parameters such as speed of the roll play a significant role in establishing the process in safe and stable operating mode. The coating process can also be affected by various defects such as uncontrolled

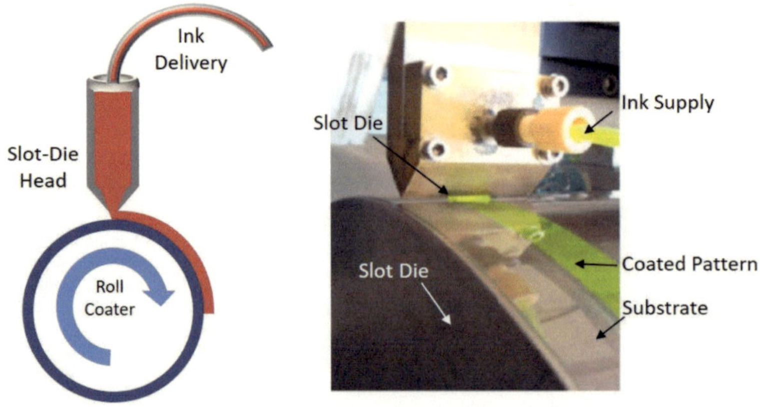

Figure 6.
Slot-die coater schematic and slot-die coater in operation diagram.

meniscus, resulting in dripping out of the solution, air bubbles entraps, and ribbing. Proper control at the start-up and shut-down cycles is highly demanding, as minor deviations could lead to deposition of the materials at unwanted positions on the substrate. Proper tuning of the material and process parameters is needed; otherwise, it would result in the wastage of the coating solution and also compromising on the shape of the patterns. This could also significantly affect the thin film quality and result in deviating from device real dimensions [17, 37, 38].

Materials' properties such as viscosity and surface tension along with the system parameters, i.e., slot size and gap, standoff distance, and decreased dip lip length, reduce the dimension of the printing jet. Proper adjustment of these parameters leads to shortening the trial time required to reach the steady-state conditions [38]. Despite the high-speed coating capabilities, the challenges involved in reaching stable operating conditions make the process less attractive than other printing systems and are adopted seldomly in the manufacturing of printed electronic devices.

2.2 Contact-based printing technologies

In contact-based printing techniques, prepatterned structures of the printing tools are physically brought in conformal contact to the target substrate. Similarly, micron-scale dispensing nozzles are also used for high-resolution patterning by contacting the target surface in a similar fashion as drawing. Almost all the techniques used in contact-based printing are precise and rapid; therefore, these techniques are used by the industry for the mass productions.

2.2.1 Screen printing

Screen printing remains the top priority when it comes to rapid, fast, and large area manufacturing. The technology has been using from the early developmental stages of microelectronic industry, especially for printing electrodes and interconnections. Screen printing is advantageous when compared to other printing systems as it is more versatile, and processing is simple, capable of reproducing similar structures in large batches with minimum dimensional variations. Results are duplicated by repeating the similar printing parameters and optimized solution pastes [12, 39].

Screen printing process can be established following the two different assemblies, i.e., in flatbed and installing a rotating surface [23, 40]. Flatbed systems are usually employed for low-throughput production and lab-level research activities. Whereas rotary systems are installed on a fast production platforms such as R2R, where all the solution and printing parameters are optimized first and applied afterward for high-speed production of devices. **Figure** 7 shows schematics of flatbed and rotary screen printing systems. The system setup is simple, which contains screen, squeegee, press bed, and substrates placed on the movable stage. In flatbed systems, the printable solution is applied on the screen, and a squeegee is used to coat the ink all the way over the structured mesh on the stencil. A controlled pressure ensures the dispensing of ink through the holes in the mesh, and the squeegee recollects the extra ink for the consequent layers. Flatbed screen printers are ideal for the optimization of the process and structures on a small-scale lab level research. For high-speed production, a folded screen with squeegee inside the rotating and ink filled in a tube are applied on a continuously rotating system. However, the screens for rotary screen systems are expansive and challenging to clean in case of material clogging [23, 41]. The print quality in both the approaches is affected greatly by the similar set of parameters such as solution viscosity, print speed, angle and geometry of the squeegee, standoff between screen and substrate, and mesh size [42–44]. Material properties are tuned separately to have the right viscosities and surface tension for complete dispensing through the screens.

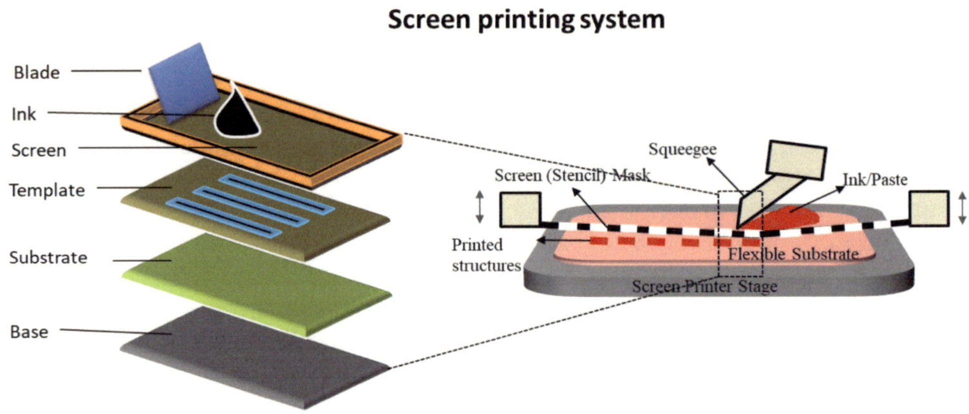

Figure 7.
Flatbed and rotary screen printing systems.

Viscosity of the pastes used with screen printing is kept higher than other conventional printing technologies in order to avoid undesired flow through the screen masks [27, 45]. Screen printing is used for both printing patterned structures and coating larger areas [46]. Pattern resolution in the ranges of 80–100 μm can be achieved after proper tuning the solution properties and optimizing the screen printing parameters. A good compromise between the surface energies of the substrate and the surface tension of the ink is desired to achieve higher resolutions [45, 47, 48]. The type, material, and strength of mesh used in the screen also contribute to the high-resolution patterning through screen printing. Different materials such as nylon, polyester, and stainless steel are used in the screen mesh. For printing stability during mass production, a screen made of stainless steel mesh with three times more in strength than conventional stainless steel mesh has also been developed [43, 49]. The possibility of printing relatively thick layers could enable printing of low-resistance structures, also with conducting polymers, by compensating the high-volume resistivity with a thicker layer [6].

Screen printing has been successfully used for demonstrating various fabricated devices. For instance, an all-screen printing has been adopted for developing thin film transistors (TFTs) [50–52]. OLED devices are also presented by exploring the different process parameters, such as viscosity and mesh count, and their effect on the printed structures [43]. An advanced screen-printing approach is adopted to develop multilayer high-density flexible electronic circuits connected through holes with embedded passive and optical devices [42]. Printing interconnect lines between discrete devices on a same substrate or tape out for data reading are usually printed with screen printing. Screen printed electrical interconnects for temperature sensor on PET substrate are reported by Shi et al. [48]. Screen printing does not require high-capital investment like many other manufacturing techniques, and setup can be established at lower installation costs. A high-speed production line can be established by assembling rotary printing setup along with supplemental methods such as inkjet, vapor deposition, and laser/flash photonic sintering tools [49, 53]. Although the advantages and attractions using screen printing are higher, there are also some challenges that need to be addressed. Printing multiple layered structures, higher wet thicknesses, and exposure of the ink to ambient environment while remaining on the screen bed present more risks for developing repeatable and reliable electronic devices. The rapid evaporation of solvents and surfactants from the printing paste, when the system is idle and ink applied on the screen lead to blockage of the screen masks [54]. Therefore, proper tuning of the material properties and developing reliable procedures are highly desired for developing a repeatable printing recipe.

2.2.2 Gravure printing

Gravure printing is the most prominent and representative technique in the contact-based printing category. Structures are transferred through a prepatterned surface, where the solvent is deposited on the target surface upon contacting. The engraved structures are designed in cylindrical shaped objects, and substrate controlled through moving rolls of the system represents a typical R2R process. The gravure printing tools consist of a large cylinder electroplated with copper and engraved with microcells, as shown in **Figure 8**. The microcells are engraved by either using electromechanical means or using laser [23, 26, 41, 55]. The physical contact between the printing surface and the substrate leads to wear and tear of the engraved structure; therefore, chrome electroplating is performed for protecting it from deterioration. The size of engraved microcells is responsible for ink pickup from the reservoir lying beneath the cylinder or filled with dispensing nozzles from the top. The extra ink is removed from the surface using a doctor blade to avoid cross contamination or unwanted deposition of ink on the substrate.

Surface properties of the substrate and rheological properties of ink are tuned to promote the efficient deposition. Capillary action of the ink plays a significant role in complete transfer of the ink from microcells of the engraved cylinder to the substrate. Pressure on the impression cylinder is also properly controlled to expedite the ink transfer and to reduce the deterioration of contacting surfaces. Width and depth ratio of the microcells in the engraved cylinder also plays a significant role in manipulating the ink transfer [41]. Viscosity and surface tension of the desired material solution have to be in acceptable ranges to prevent the bleeding out of the solution from the microcells. The optimal viscosities help in rapid prototyping by increasing the printing speed and allow full emptying of the ink from engraved microcells [56].

An advanced version of gravure printing, i.e., gravure offset uses an extra elastic blanket to avoid potential risks occurring from the deteriorations of the contacting surfaces, as shown in **Figure 8**. The elastic blanket serves as an intermediate step between the contacting surfaces, where the ink from engraved microcells is picked up by the blanket and transferred finally on the target substrate. This avoids the direct interface between the patterned cylinder and the substrate. Few of the printing parameters affecting significantly the print quality are speed, pressure, and blanket's dimensions plus thickness. Surface properties of the blanket and its thickness combined with the speed are more dominant factors that control and enhance the efficient transfer of the ink [16, 41, 57]. The limited time of contact and

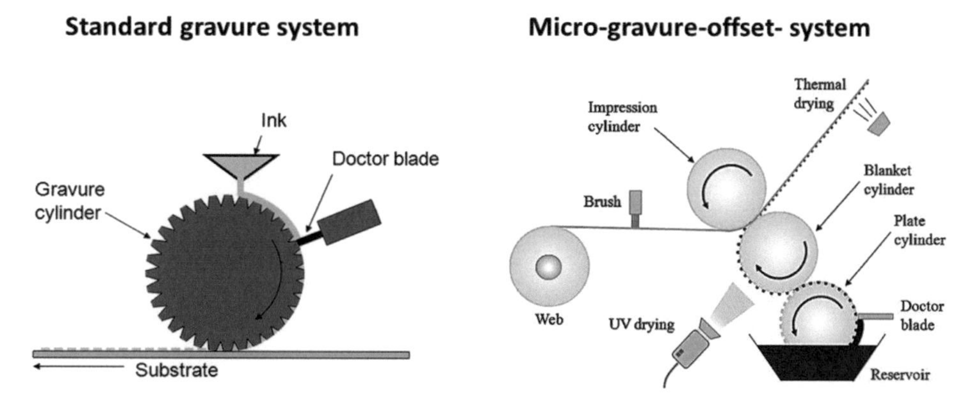

Figure 8.
Schematic of a typical standard and microgravure-offset printing process.

higher speed also enable high-resolution patterning on the substrate and increase reliability of the system. Reliability of gravure offset is more critical for assembling in a high-speed production line of printed electronics on rollable substrates [58]. Combinations of various manipulating forces such as adhesive force between the blanket and the ink, cohesive force within the ink when on the blanket, adhesive force between the ink and the gravure, and adhesive force between the ink and the target substrate are of particular importance to control and tune for efficient transfer of ink. A complete dispensation of the ink is desired as a minor mismatch or open hole within the pickup ink can lead to incomplete printing of the structures [16, 59]. Speed of the roll is also a main contributing factor to optimize for complete transfer of the ink, and a uniform impression pressure in suitable ranges would increase the uniformity of the print edges and product yield. Despite the advantages offered by using an intermediate medium of transfer, gravure offset also poses some serious challenges. For instance, the lifespan of the blanket is of serious concern, as continuous absorbance of the ink leads to saturation of the surface and needs to be changed quite often. Prolonged use of the blanket greatly affects the resolution of the printed patterns on target substrates as it reduces temporarily the ink viscosity, which results in spread out of the ink during the setting process [60]. Therefore, a proper and timely maintenance of the printing tools, especially the blanket, is needed to avoid undesired printing structures.

2.2.3 Flexographic printing

Flexographic printing ensures high-speed printing and produces high-resolution patterned structures as compared to gravure and gravure-offset printing approaches [23]. A rubber- or polymer-based plate with elevated patterns on the surface and developed through photolithography is used in flexographic printing. The plate is attached to the printing cylinder as shown in **Figure 9**. A wide variety of ink including but not limited to solvent-based, wafer-based, UV-curable inks and two parts chemically curing inks, etc. can be processed to pattern high-resolution structures on target surfaces [18, 61, 62]. The Anilox cylinder picks up the ink from the reservoir, and upon contacting the inked areas with the plate cylinder, it transfers the ink and prints on the running substrate between plate and impression cylinders as shown in **Figure 9**. Transfer of the ink is more efficient in flexographic printing and results in very thin patterns with sharper edges. Amount of ink pick up is controlled predominantly by the Anilox roll, where the size and frequency of the engraved cells

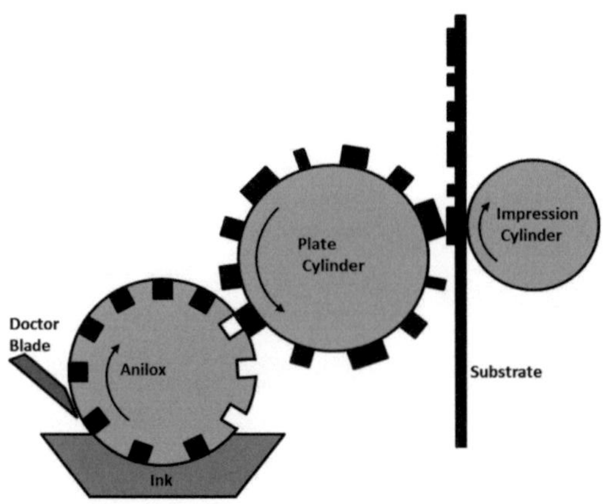

Figure 9.
Flexographic printing schematic diagram.

are responsible indirectly. A proper balance between the solution properties such as nanoparticles mixing ratios plus the carrier fluids is essentially required for filling the Anilox engraved cells. Relatively higher-solution concentrations are required within the specific range of viscosities to achieve good resolution patterns [18, 61, 62]. Typical resolution obtained with flexographic printing is in the range of 50–100 μm; however, further higher resolutions down to ~20 μm could be made possible by proper tuning the solution properties and process the parameters [18, 61, 62]. Film quality printed with flexography is uniform as compared to other competitive printing technologies [60]. Film instability and dewetting of the printing plates cause many defects such as open lines, overlapped lines, and edge waviness. Controlling the load pressure and cell aspect ratio is very critically important to avoid these issues, especially in the case of targeting high-resolution printing and high-end devices. Maintenance plus observation of the engraved cells is very important, as the blockade or erosion of any of the cells could lead to discontinuous printed patterns. An acceptable margin in terms of resolution needs to be allowed, as the pattern dimensions could vary as a result of pressure from impression cylinder on the flexible or polymer plate [18, 27, 61–63]. An optimum range of width and thickness is needed for the printed patterns to decrease the ohmic losses and also increase the efficiency of the printed devices [47]. For thick film deposition through flexographic, several printing passes with similar parameters are required, which also minimize sheet resistance. Repeating the same procedure needs proper alignment of the equipment for subsequent layers, which adds to the complexities of the system [64]. The current technology limits the highly desirable features such as high-switching speed and reduced supply voltage that are needed for many applications. These limitations result in degraded device parameters such as charge-carrier mobility, parasitic capacitances, and overlay precision registration accuracy [65]. Challenges to overcome for fine patterning are surface irregularities and pores, nonuniform films, ragged lines, and nonavailability of suitable functional materials [66].

2.2.4 Microcontact printing (μCP)

Microcontact printing is a special type of contact-based printing approach, where an inked surface is brought in conformal contact and transfers the patterns on target surface. The contact is controlled through micromanipulation, and surface conditions are set to release and receive the ink consequently. A conformal contact of prepatterned elastomeric stamp with precise control and alignment to the target surface on micron scale is key requirement for successful transfer of the structures. A master mold is developed using conventional microfabrication or photolithography techniques, and multiple copies of the stamp with desired structures are reproduces. A moldable and elastomeric material is usually used to develop the stamp, which can easy be casted into the master mold and delaminated without causing any degradation of the microscale structures on stamp surface [3, 19, 67, 68]. Poly(dimthylsiloxane) (PDMS) is the frequently used elastomer due to its extraordinary properties as compared to other elastomers such as polyurethanes, polyimides, and cross-linked Novolac resin. PDMS has few distinguishing properties such as conformability to larger areas, deformable to mount on nonplanar surfaces, elasticity for easy release, isotropic, and optically transparent. Low-surface free energy, chemically inertness, and durability for multiple uses make PDMS an attractive candidate for the purpose of using it as a stamp [19, 68]. Microcontact is an effective rapid prototyping technique for preparing the substrates and patterning a wide range of materials that are sensitive to light and chemical etchants. The lower surface energies offered by PDMS stamp due to elasticity of siloxane chain and the lower intermolecular forces between the methyl groups promote the peeling

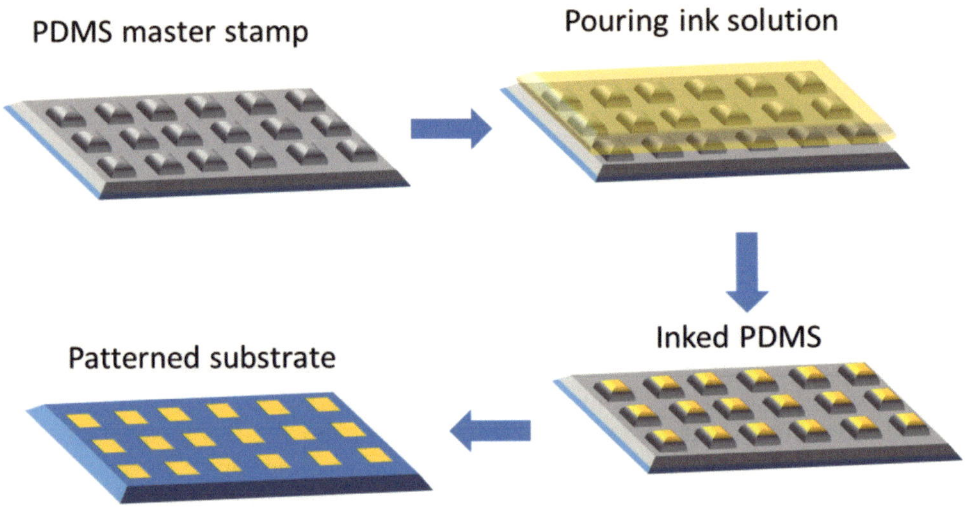

Figure 10.
Microcontact printing (μCP) steps during processing.

and stamping capabilities, which are ideal for microcontact printing. Surface energies of the stamp and target surface play a significant role in efficient transfer of the ink. For instance, a higher energy of PDMS stamp is desired for the pickup of ink from the donating surface, whereas higher energy than the PDMS stamp is desired for the receiving surface to complete detachment of the ink. To avoid collapsing of the stamp during peeling or capillary action during inking, a specific ratio of the height to width of the features on the stamp is required [69]. **Figure 10** shows schematic of the process flow of typical microcontact printing approach.

Robustness of the stamp by having sufficient flexibility and mechanical strength for maintaining the dimensional integrity of the printable structures is of great importance. Elastomeric properties of the stamp material enhance the efficient ink delivery by providing a good interface between stamp, ink, and target substrate. Submicron scale is challenging to reach as due to the elastomeric nature of stamp, it tends to collapse, and noncontact areas of the stamp also interface with the substrate leading to uneven printing. Besides the printing tools and processing condition, chemical composition of the ink also contributes to the microcontact process. For instance, polar molecules are challenging to print with stamp because of the hydrophobicity of the PDMS. Therefore, treatment at right conditions of the PDMS stamp is central for stamping such type of materials [19]. Similar challenge faced by the flexographic technique is microcontact printing, as diffusion of molecules in the unpatterned areas of the stamp broadens the feature size and thus results in unwanted printing or stamping of the materials besides the intended structures [27, 70, 71]. Repeated use of the same stamp results in swelling due to the absorption of the ink and results in deterioration of the micron-scale structures on the stamp. A proper check and observation of the stamp are needed because the continuous use can cause pairing, buckling, or roof collapse of the microstructured patterns upon physical contact. Therefore, for reproducing the similar structures on target substrate, full operating conditions need to be explored considering especially the applied pressure, peeling the stamp from master mold, polarity of the molecules, and so on to guarantee uniform printing [69–71].

2.2.5 Nanoimprinting (NI)

Nanoimprinting, as the name suggests, is used to produce structures at nanoscale by using an imprint approach. NI uses mechanical and physical deformation of wet

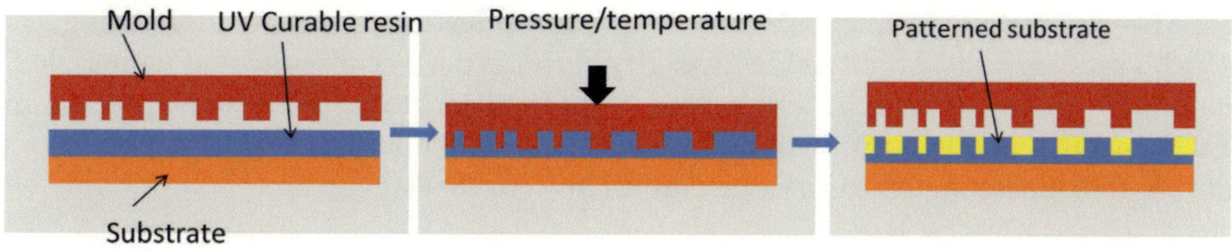

Figure 11.
Nanoimprint steps for patterning the substrate.

layers through molding accompanied by different thermal procedures. The NI operating principles are quite straight forward, as shown in **Figure 11**. As against micro-contact printing, NI uses a mold having nanoscale structures developed through standard clean room processes and is pressed against a uniformly coated wet surface at controlled pressure and temperature. A thin-residual layer of polymeric material is intentionally left underneath the mold protrusions and acts as a soft cushioning layer that prevents direct impact of the hard mold on the substrate and effectively protects the delicate nano-scale features on the mold surface [72]. Resist filling and demold characteristics are the two primary and critical processing steps affecting the print quality and throughput. Controlling the pressure precisely is central during the demolding process, which helps in maintaining the imprinted patterns at the desired dimensions [15]. Various approaches are adopted for executing NI process such as thermal, ultraviolet (UV), step and flash, and roller imprinting. They are all selected on the basis of type of material and processing conditions. In UV-NI process, transparent medium is required for allowing the UV light to penetrate through and in-situ sintering of the imprinted structures. UV-IN is thus considered the most reliable and robust approach among the practiced NI procedures. Common materials for the molding in UV-IN are quartz and silica, which are molded by using very high resolution electron beam lithography techniques [72–75].

Compared to the quartz and mechanically rigid, polymer-based molds are advantageous as they can replicate nanostructures over larger areas. Besides, the material cost is lower, and it can be used to develop manufacturing platform at depreciated costs. The mechanical molding makes the process a bit challenging embarking into new challenges compared to the traditional manufacturing processes [76]. The spatial confinement of solution-based materials in nanoscale ranges gives rise to drastic changes in the physical, electrical, and chemical properties as compared to the patterned structures in micro or macrostructures. NI offers several advantages such as very high resolution patterning, high pattern transfer fidelity, 3D patterning, covering large areas, reduced fabrication steps, high throughput, and lower processing cost. However, the challenges faced by NI process overshadow the attractions and keep it restricted for a special case of uses. The overlay alignment, template fabrication, defect control, yield, and seeking a suitable application with a good payback are few of the critical challenges faced by NI process [15, 22, 72, 75, 77]. Probability of defect density is higher, and the mechanical and physical inabilities to withstand the uneven pressures on the molding masks lead to the lateral and vertical collapse of the nanostructures. This adds to the reproducing cast, and attaining repeatable structures with similar dimensions are challenging due to the less margin in dimensional variability. The thermal-based NI involves in-situ heating while applying the mask on the wet film, and slight variations in the temperature or thermal mismatch between the mask and the printed materials could lead to deterioration of the pattern structures on the mask. Time required, i.e., 10–15 min per replication for heating and cooling cycles, is longer than other soft printing techniques [74]. The thermal budget of about 125°C is challenging for some of the plastic substrates with lower glass transition temperature, which can create dimensional instabilities. Several electronic devices have been presented by different research groups following the NI manufacturing. The NI process is ideal for a single-layer structure, as perfect alignment at nanoscale on multilayer devices is a challenging task. As a final goal to transfer the NI process on a R2R manufacturing, introducing multilayer steps and misalignment of the stamp during the imprint needs to be taken care of [69, 70, 78–80].

2.2.6 Transfer printing

Transfer printing is relatively a new technique to fabricate flexible electronics through physical transfer of prefabricated structures using a stamp [81]. Microstructures in the shape of wires or membranes are developed using standard photolithography processes in clean room, etched underway, and then used a stamp to pick and stamp on a target surface. Transfer printing can be executed in two ways such as direct transfer and stamp-assisted transfer. In direct transfer, an adhesive coating is performed on the receiving surface, and the donor wafer is directly contacted with it. After releasing, the structures are transferred to flipped surfaces on the target surface. Whereas, stamp-assisted transfer is completed in two steps: in first, the microstructures from donor substrate are picked up the stamp, and the stamp is contacted with the target substrate, thus transferring the structures with the top processed surface facing upward. The elastomeric stamp usually developed by using a PDMS is used, where the viscoelastic properties of the stamp are exploited to detach the fabricated structures from donor substrates and place them deterministically on a secondary substrate [82–86]. This technique is ideal for developing high-speed electronic devices on unconventional substrates, by merging both organic- and inorganic-based materials. The heterogeneous integration of these dissimilar materials helps in reducing the cost and maintaining the reliability of developed electronic circuits and systems [87]. The lower cost comes from the large area processing units such as inkjet printing for the electrodes and interconnections as well as insulator layers, where the high-mobility semiconductor layers are integrated into discrete circuit or into full circuit through stamp-assisted transfer technology. This is an effective approach toward manufacturing of high-end devices on larger areas as the state-of-the-art processing of electronic grade silicon, and other compound semiconductors in clean rooms constitute high level of purity, surface smoothness, control over crystallinity, doping levels, and types, resulting in higher carrier mobilities.

The structures are developed and finished in clean room processes and under-etched afterward to release and be ready for the PDMS transfer. **Figure 12** shows schematics of the processing steps involved in transfer printing approach. Here, the structures are shown in the shape of wires, which are released from the underlying layers using the relevant etchants. Tethered points are designed as per the dimensional requirements and are sufficient enough to keep intact the released structures with the donor wafer before contacting the PDMS stamp. The dimension of tethered points is essential and useful for efficient transfer, especially for structures with dimensions in the micron scale. The conformal contact of soft elastomeric stamp having surface activated temporarily through oxygen plasma with the microwires or microribbons guarantees the efficient pickup. The wires attach to the stamp surface and, when peeled back, retrieve the microstructures with fast speed, enhancing the kinetic control of adhesion [88]. The rate-dependent adhesion and printing of the solid structures with high-peel velocity (typically 10 cm/s) and low-stamping velocity (~1 mm/s), respectively, have been investigated [89]. The mechanics of kinetic dependence of switching of adhesion between the microstructures and the stamp has its origin in the viscoelastic response of the elastomeric materials, i.e., PDMS. Adhesiveless stamping like this is very valuable for wafer-based microstructure printing to operate it from moderate to high temperatures [82].

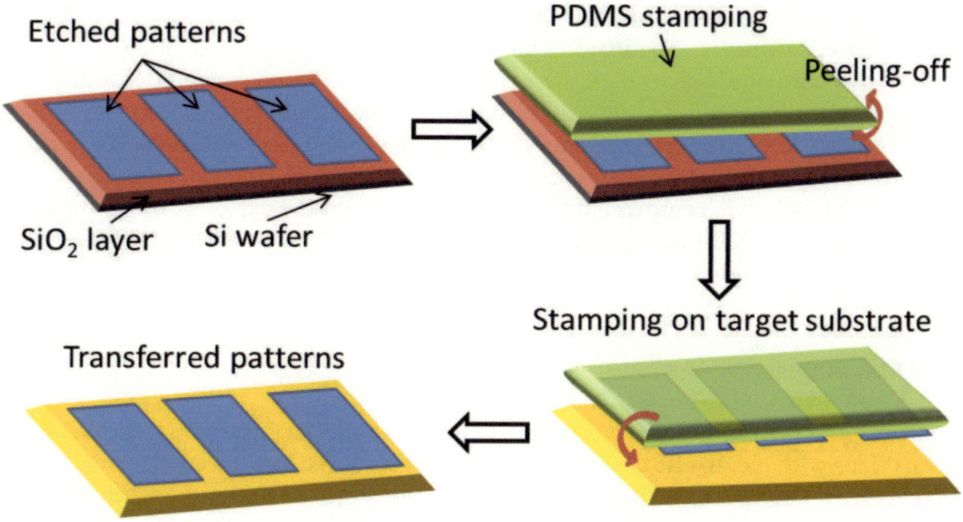

Figure 12.
Conformal contact of polymeric stamp with the Si patterns, pickup of patterns by peeling off the stamp, and patterns transferred to final substrate.

3. Mass production perspective

Developing printing technologies is motivated by the development of fast and efficient production line by assembling different manufacturing units, substrate treatments, and sintering after fabrication, as shown in **Figure 13**. The highly optimized techniques on lab level are merged together to develop a single manufacturing platform. The scope of such developments is rendered from the high-speed text printing-based technology. A similar approach is foreseen to establish for batch manufacturing of electronic components and systems on larger areas as much higher speeds. Printing processes matured at lab level need to be transferred to large-scale fast production lines with the same level of performance. However, assembling different printing techniques into a single production platform inline is a challenging task, as all the processing parameters and conditions need to be properly tuned and need to be in close ranges of the process conditions, especially when the substrate is moving at higher speeds as 5–50 m/min. Therefore, investigation of the optimal and matching processing conditions is desired for the new arrangement, which is not at all obvious, especially when considering that there are boundary conditions

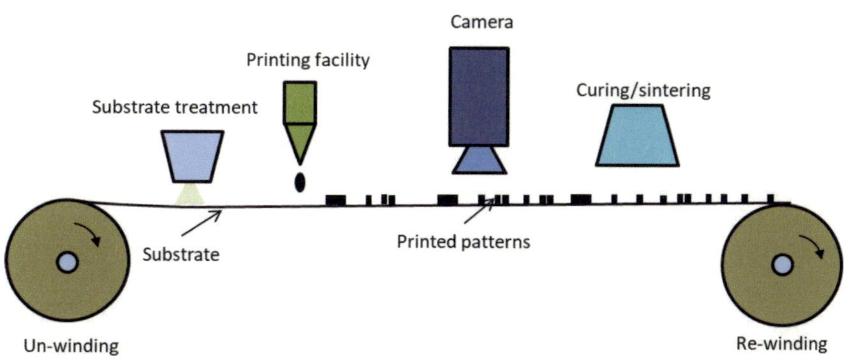

Figure 13.
Schematic illustration of typical roll-to-roll system where different deposition, patterning, and sintering modules are installed.

(i.e., materials, solvents, multilayer processing, overlay registration accuracy, drying temperature, speed, etc.) involved in fast R2R processing [20, 23, 53, 90].

R2R as a commonly shared platform has the potential for a continuous and high-throughput process for deposition of diverse materials on large substrate rolls (often called "web") [53, 90]. Besides the instrumentation and hardware for control system, R2R line is equipped with several rollers over which the web (flexible substrates) passes with controlled tension. As described in Section 3, these webs are the backbone of a R2R system and should be accurately controlled during passage through different rollers and processing sections.

4. 3D packaging

Packaging is an important step during electronic fabrication, which enables user to interface with electronic devices, circuits, and systems. In printed electronics, devices are relatively large size as compared to conventional technologies and easy to handle. However, they need packaging to protect from the ambient environment such as humidity, light, and temperature.

On the other hand, the device itself needs protection from the user touches as it can damage the thin films and patterns. 3D printing (also called additive manufacturing) is a manufacturing technology, which is based on imposing the material layers to create the 3D objects. A 3D object is fabricated through melting filament material with controlled temperature and flow rate in combination with X, Y, and Z axis control, as shown in **Figure 14**. The object is design in CAD tool, i.e., AutoCAD or any other tool that can create 3D structures and converted into printer supported file format. The file is then loaded into the printer to create the object in 3D form. There are several ways to create a 3D object, which defines the types of 3D printers.

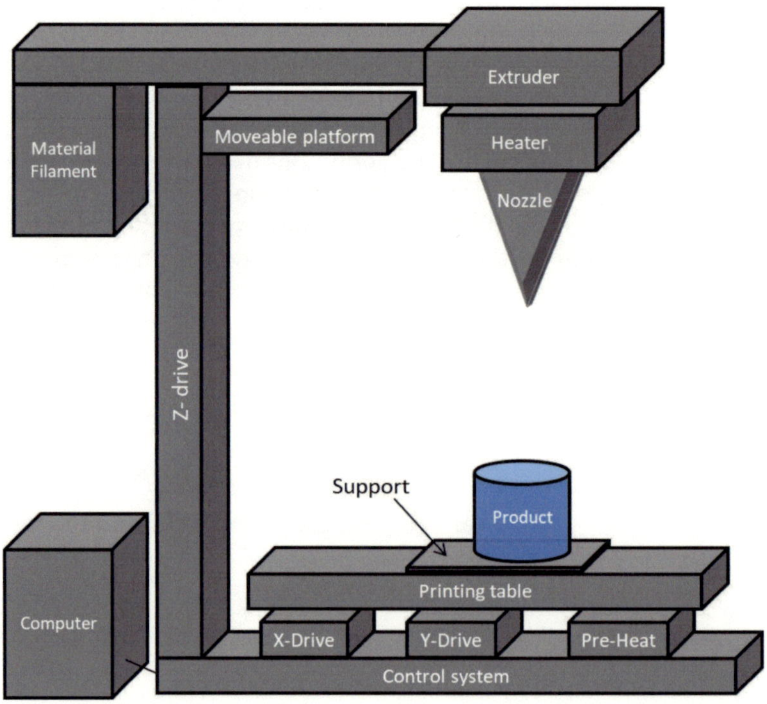

Figure 14.
3D printing system schematic diagram.

4.1 Fused deposition modeling

The development of 3D object is made by either microdrops or melt near-field electrospinning of melted thermoplastics of consecutive layers, which solidifies after a certain time. Commonly used filament materials are PVA, PLA, ABS, nylon, and some composites. This 3D printing technology is often used in the rapid prototyping objects.

4.2 Stereolithography

Stereolithography (SLA) uses a UV laser instead of melting the filament through heater and making microdrops. An SLA printer uses two mirrors in combination with UV laser, known as galvanometers, positioned on the X-axis and on the Y-axis. Both galvanometers rapidly aim a laser beam across a vat of resin, the area under light beam selectively curing and solidifying a cross section of the object inside this build area, building it up layer by layer and forming a 3D structure.

4.3 Digital light processing (DLP)

This 3D printing technology is almost the same as stereolithography, the only difference that instead of laser and two mirrors, DLP uses image of the 3D object to make one layer and repeat until the job is finished. DLP is much faster than SLA as it uses digital image array to produce a structure in the vertical sequential order frame by frame. The digital image of the target object is consisted of small rectangles called voxels. This 3D printing technology is based on selective laser sintering (SLS), and commonly used materials are thermoplastic powders (Nylon 6, Nylon 11, and Nylon 12). Potential applications of this technology are functional parts, complex ducting (hollow designs), and low run part production. This technology can be used for strong and elastic mechanical property parts and also for the complex geometry printing.

4.4 Selective laser sintering (SLS)

SLS 3D technology creates an object with powder bed fusion technology and polymer powder. Working of this technology, i.e., polymer powder, is preheated to a temperature slightly below the melting point. Then, a very thin layer of the powdered material is deposited with the help of a blade normally 0.1 mm thick on the object platform. The surface is then scanned with a CO_2 laser beam, as it selectively sinters the powder and solidifies a cross section of the object according to the designed geometry. Same as SLA 3D technology, the laser is precisely focused on to the correct location with the help of two galvos. Once the entire cross-sectional area of the object is scanned that creates one layer of the object, the build platform will move down one layer thickness in height to make the next step happen. The powder recoating blade deposits a new layer of the powder on the top of prescanned layer, and the laser will sinter the next cross section of the object area the same as previous layer. This process is continuous until the object is created.

5. Conclusion

Commonly used and advanced printed electronic technologies were discussed briefly in this chapter that covers almost all the technologies to fabricate electronic devices, circuits, and systems. Although the printed electronic technologies come with their limitations of scalability, mass production, and life time, it is not perfect replacement of the conventional electronic technology; however, it allows free design, rapid prototyping, and unlimited application areas, especially low-temperature fabrication. Printed electronic technology benefits from new printing techniques, solution-based materials, and the combination of other manufacturing processes. Printing technology has a prominent impact on the electronic industry, such as flexibility and biocompatibility; otherwise, it was impossible to achieve with the conventional techniques and materials. Different printing technologies, processing requirements, operation, materials, and fabrication limitations were highlighted. The possibility of combining printing technologies to enable mass production of the devices, circuits, and systems, i.e., R2R system, was also presented. Moreover, it was discussed that 3D printing plays an important role in the packaging and test platform fabrication for the printed electronics. At present stage, printed electronics suffers from limitations such as operating frequency, scalability, mass production, shelf life, and robustness. However, as the research is continuous in the field, breakthroughs are expected in near future to meet all these challenges.

Acknowledgements

This work was supported by NPRP from the Qatar National Research Fund (a member of Qatar 846 Foundation) under Grant NPRP10-0201-170315 and NPRP11S-0110-180246. The publication charges were supported by Qatar National Library (QNL). The findings herein reflect the work and are solely the responsibility of the authors.

Conflict of interest

Authors declare no conflict of interests.

Author details

Saleem Khan[1*], Shawkat Ali[1,2] and Amine Bermak[1]

1 College of Science and Engineering, Hamad Bin Khalifa University, Doha, Qatar

2 Electrical Engineering, National University of Computer and Emerging Sciences (FAST-NU), Islamabad, Pakistan

*Address all correspondence to: sakhan3@hbku.edu.qa

References

[1] Khan S, Lorenzelli L, Dahiya RS. IEEE Sensors Journal. 2015;**15**(6):3164

[2] Khan S, Briand D. Flexible and Printed Electronics. 2019;**4**(1):015002

[3] Michel B, Bernard A, Bietsch A, Delamarche E, Geissler M, Juncker D, et al. IBM Journal of Research and Development. 2001;**45**(5):697

[4] Ali S, Bae J, Lee CH, Choi KH, Doh YH. Sensors and Actuators B: Chemical. 2015;**220**:634

[5] Ali S, Bae J, Choi KH, Lee CH, Doh YH, Shin S, et al. Organic Electronics. 2015;**17**:121

[6] Tobjörk D, Österbacka R. Advanced Materials. 2011;**23**(17):1935

[7] Ali S, Bae J, Lee CH. Applied Physics A. 2015;**119**(4):1499

[8] Oliveira J, Correia V, Castro H, Martins P, Lanceros-Mendez S. Additive Manufacturing. 2018;**21**:269

[9] Ali S, Hassan A, Bae J, Lee CH, Kim J. Langmuir. 2016;**32**(44):11432

[10] Ali S, Bae J, Lee CH, Choi KH, Doh YH. Organic Electronics. 2015;**25**:225

[11] Ali S, Bae J, Lee CH. Electronic Materials Letters. 2016;**12**(2):270

[12] Ali S, Hassan A, Khan S, Bermak A. SN Applied Sciences. 2019;**1**(7):744

[13] Khan S, Lorenzelli L. Smart Materials and Structures. 2017;**26**(8):083001

[14] Khan S, Lorenzelli L, Dahiya R. IEEE Journal of the Electron Devices Society. 2016;**4**(4):189

[15] Radha B, Lim SH, Saifullah MSM, Kulkarni GU. Scientific Reports. 2013;**3**:1078

[16] Kang HW, Sung HJ, Lee TM, Kim DS, Kim CJ. Journal of Micromechanics and Microengineering. 2009;**19**(1):015025

[17] Sandström A, Dam HF, Krebs FC, Edman L. Nature Communications. 2012;**3**:1002

[18] Deganello D, Cherry JA, Gethin DT, Claypole TC. Thin Solid Films. 2010;**518**(21):6113

[19] Kaufmann T, Ravoo BJ. Polymer Chemistry. 2010;**1**(4):371

[20] Jung M, Kim J, Noh J, Lim N, Lim C, Lee G, et al. IEEE Transactions on Electron Devices. 2010;**57**(3):571

[21] Ali S, Hassan A, Hassan G, Bae J, Lee CH. Carbon. 2016;**105**:23

[22] Pease RF, Chou SY. IEEE Proceed. 2008;**96**(2):248

[23] Søndergaard RR, Hösel M, Krebs FC. Journal of Polymer Science, Part B: Polymer Physics. 2013;**51**(1):16

[24] Leenen MAM, Arning V, Thiem H, Steiger J, Anselmann R. Physica Status Solidi (a). 2009;**206**(4):588

[25] Ali S, Hassan A, Hassan G, Eun C-H, Bae J, Lee CH, et al. Scientific Reports. 2018;**8**(1):5920

[26] Subramanian V, Chang JB, de la Fuente Vornbrock A, Huang DC, Jagannathan L, et al. Presented at the 34th European Solid-State Circuits Conference, 2008. ESSCIRC 2008. 2008 (unpublished)

[27] Moonen PF, Yakimets I, Huskens J. Advanced Materials. 2012;**24**(41):5526

[28] Khan S, Doh YH, Khan A, Rahman A, Choi KH, Kim DS. Current Applied Physics. 2011;**11**(1):S271

[29] Hassan G, Khan F, Hassan A, Ali S, Bae J, Lee CH. Nanotechnology. 2017;**28**(17):175402

[30] Hassan G, Bae J, Khan MU, Ali S. Materials Science and Engineering: B. 2019;**246**:1

[31] Ali S, Bae J, Lee CH. Optics Express. 2015;**23**(24):30583

[32] Choi KH, Khan S, Dang HW, Doh YH, Hong SJ. Japanese Journal of Applied Physics. 2010;**49**(5S1):05EC08

[33] Khan S, Nguyen TP, Lubej M, Thiery L, Vairac P, Briand D. Microelectronic Engineering. 2018;**194**:71

[34] Singh M, Haverinen HM, Dhagat P, Jabbour GE. Advanced Materials. 2010;**22**(6):673

[35] Lin CF, Wong DSH, Liu TJ, Wu PY. Advances in Polymer Technology. 2013;**32**(S1):E249

[36] Lin CF, Wong DSH, Liu TJ, Wu PY. Advances in Polymer Technology. 2010;**29**(1):31

[37] Hu L, Kim HS, Lee J-Y, Peumans P, Cui Y. ACS Nano. 2010;**4**(5):2955

[38] Chang Y-R, Lin CF, Liu TJ. Polymer Engineering and Science. 2009;**49**(6):1158

[39] Ali S, Khan S, Bermak A. Sensors (Basel). 2019;**19**(5):1197

[40] Khan S, Tinku S, Lorenzelli L, Dahiya RS. IEEE Sensors Journal. 2015;**15**(6):3146

[41] Clark DA. Major trends in gravure printed electronics. 2010

[42] Turunen R, Numakura D, Nakayama M, Kawasaki H. Presented at the IPC Printed Circuit Expo/APEX 2008 (unpublished)

[43] Ghassan J, Rachel ER, Peyghambarian N. IEEE Journal of Selected Topics in Quantum Electronics. 2001;**7**(5):769

[44] Soukup R, Hamacek A, Reboun J. Presented at the 35th Intl. Spring Seminar on Electronics Technology (ISSE). 2012 (unpublished)

[45] Liang TX, Sun WZ, Wang LD, Wang YH, Li HD. IEEE Transactions on Components Packaging and Manufacturing Technology. 1996;**19**(2):423

[46] Khan S, Dang W, Lorenzelli L, Dahiya R. IEEE Transactions on Semiconductor Manufacturing. 2015;**28**(4):486

[47] Siden J, Nilsson HE. Presented at the IEEE Intl. Symp. Antennas Propag. Soc. 2007 (unpublished)

[48] Shi CWP, Shan X, Tarapata G, Jachowicz R, Weremczuk J, Hui HT. Microsystem Technologies. 2011;**17**(4):661

[49] Sano Y. CONVERTECH. Japan: Converting Technical Institute; 2012. p. 1

[50] Kwack YJ, Choi WS. Journal of the Korean Physical Society. 2011;**59**:3410

[51] Noguchi Y, Sekitani T, Someya T. Applied Physics Letters. 2007;**91**(13):133502

[52] El Houti EJI, Thierry T, Moliton JP. Micro and Nanosystems. 2009;**1**(1):46

[53] Schwartz E. Roll to roll processing for flexible electronics. Cornell University. 11 May 2006:11

[54] Krebs FC, Jørgensen M, Norrman K, Hagemann O, Alstrup J, Nielsen TD, et al. Solar Energy Materials & Solar Cells. 2009;**93**(4):422

[55] Pudas M, Halonen N, Granat P, Vähäkangas J. Progress in Organic Coating. 2005;**54**(4):310

[56] Yang J, Vak D, Clark N, Subbiah J, Wong WWH, Jones DJ, et al. Solar Energy Materials & Solar Cells. 2013;**109**:47

[57] Lee TM, Lee SH, Noh JH, Kim DS, Chun S. Journal of Micromechanics and Microengineering. 2010;**20**(12):125026

[58] Lee TM, Noh JH, Kim I, Kim DS, Chun S. Journal of Applied Physics. 2010;**108**(10):102802

[59] Choi N, Wee H, Nam S, Lavelle J, Hatalis M. Microelectronic Engineering. 2012;**91**:93

[60] Yan H, Chen Z, Zheng Y, Newman C, Quinn JR, Dötz F, et al. Nature. 2009;**457**(7230):679

[61] Deganello D, Cherry JA, Gethin DT, Claypole TC, Gethin DT, Claypole TC. Thin Solid Films. 2012;**520**(6):2233

[62] Maksud MI, Yusof MS, Jamil A, Mahadi M. International Journal of Integrated Engineering. 2014;**5**(3):679

[63] Lo C-Y, Keinänen JH, Huttunen O-H, Petäjä J, Hast J, Maaninen A, et al. Microelectronic Engineering. 2009;**86**(4):979

[64] Kwak MK, Shin KH, Yoon EY, Suh KY. Journal of Colloid and Interface Science. 2010;**343**(1):301

[65] Kempa H, Hambsch M, Reuter K, Stanel M, Schmidt GC, Meier B, et al. IEEE Transactions on Electron Devices. 2011;**58**(8):2765

[66] Lee HK, Joyce MK, Fleming PD. Journal of Imaging Science and Technology. 2005;**49**(1):54

[67] Kim P, Kwon KW, Park MC, Lee SH, Kim SM, Suh KY. BioChip Journal. 2008;**2**(1):1

[68] Rogers JA, Nuzzo RG. Materials Today. 2005;**8**(2):50

[69] Ruiz SA, Chen CS. Soft Matter. 2007;**3**(2):168

[70] Perl A, Reinhoudt DN, Huskens J. Advanced. Materials. 2009;**21**(22):2257

[71] Quist AP, Pavlovic E, Oscarsson S. Analytical and Bioanalytical Chemistry. 2005;**381**(3):591

[72] Guo LJ. Advanced Materials. 2007;**19**(4):495

[73] Nagato K. Polymers. 2014;**6**(3):604

[74] Wolfe DB, Love JC, Whitesides GM. Encyclopedia of Nanoscience and Nanotechnology. New York: Marcel Dekker, Inc; 2004. pp. 2657-2666

[75] Gilles S, Meier M, Prömpers M, van der Hart A, Kügeler C, Offenhäusser A, et al. Microelectronic Engineering. 2009;**86**(4):661

[76] Schift H. Journal of Vacuum Science and Technology B. 2008;**26**(2):458

[77] Jiang L, Chi L. Strategies for High Resolution Patterning of Conducting Polymers. Rijeka: IntechOpen; 2010. p. 656

[78] Auner C, Palfinger U, Gold H, Kraxner J, Haase A, Haber T, et al. Organic Electronics. 2010;**11**(4):552

[79] Li B, Zhang J, Ge H. Applied Physics A. 2013;**110**(1):123

[80] Moonen PF, Vratzov B, Smaal WTT, Kjellander BKC, Gelinck GH, Meinders ER, et al. Organic Electronics. 2012;**13**(12):3004

[81] Khan S, Lorenzelli L, Dahiya R. Semiconductor Science and Technology. 2017;**32**(8):085013

[82] Meitl MA, Zhu ZT, Kumar V, Lee KJ, Feng X, Huang YY, et al. Nature Materials. 2006;**5**(1):33

[83] Zhang K, Seo JH, Zhou W, Ma Z. Journal of Physics D: Applied Physics. 2012;**45**(14):143001

[84] Dahiya RS, Adami A, Collini C, Lorenzelli L. Microelectronic Engineering. 2012;**98**:502

[85] Oh TY, Jeong SW, Chang S, Choi K, Ha HJ, Ju BK. Applied Physics Letters. 2013;**102**(2):021106

[86] Khan S, Dahiya R, Lorenzelli L. Presented at the ESSDERC 2014, the 44th European Solid-State Device Research Conference; Venice, Italy. 2014 (unpublished)

[87] Khan S, Yogeswaran N, Taube W, Lorenzelli L, Dahiya R. Journal of Micromechanics and Microengineering. 2015;**25**(12):125019

[88] Mack S, Meitl MA, Baca AJ, Zhu ZT, Rogers JA. Applied Physics Letters. 2006;**88**(21):213101

[89] Carlson A, Bowen AM, Huang Y, Nuzzo RG, Rogers JA. Advanced Materials. 2012;**24**(39):5284

[90] Huang YY, Chen J, Yin Z, Xiong Y. IEEE Transactions on Components, Packaging, and Manufacturing Technology. 2011;**1**(9):1368

Chapter 8

Plastic Inorganic Semiconductors for Flexible Electronics

Tian-Ran Wei, Heyang Chen, Xun Shi and Lidong Chen

Abstract

Featured with bendability and deformability, smartness and lightness, flexible materials and devices have wide applications in electronics, optoelectronics, and energy utilization. The key for flexible electronics is the integration of flexibility and decent electrical performance of semiconductors. It has long been realized that high-performance inorganic semiconductors are brittle, and the thinning-down-induced flexibility does not change the intrinsic brittleness. This inconvenient fact severely restricts the fabrication and service of inorganic semiconductors in flexible and deformable electronics. By contrast, flexible and soft polymers can be readily deformed but behave poorly in terms of electrical properties. Recently, Ag_2S was discovered as the room-temperature ductile inorganic semiconductor. The intrinsic flexibility and plasticity of Ag_2S are attributed to multicentered chemical bonding and solid linkage among easy slip planes. Furthermore, the electrical and thermoelectric properties of Ag_2S can be readily optimized by Se/Te alloying while the ductility is maintained, giving birth to a high-efficiency full inorganic flexible thermoelectric device. This chapter briefly reviews this big discovery, relevant backgrounds, and research advances and tries to demonstrate a clear structure-performance correlation between crystal structure/chemical bonding and mechanical/electrical properties.

Keywords: flexible electronics, plastic inorganic semiconductors, chemical bonds

1. Introduction

Flexible electronics endows electronic circuits with novel flexibility, foldability, and scalability, breaking the restriction of wafers and greatly expanding the application in various fields, e.g., flexible displays, electronic textiles, and sensory skins [1–3]. Consequently, flexible electronics has attracted widespread attention from both academic and industrial communities and has witnessed marvelous breakthroughs in terms of material development [4–5], device fabrication, packaging, and integration [6–8].

In flexible electronics, semiconductors and devices are mounted onto flexible substrates, mostly polymers [1, 9]. The key to flexible electronics is realizing both high flexibility and desired physical/chemical properties (mostly electrical performance) of the semiconductors. Due to the intrinsic softness and flexibility, polymer semiconductors have long been a popular candidate for flexible electronics [10–13]. However, carriers in organic materials are rather localized, leading to poor electrical conduction. Another way to realize "flexibility" is to process inorganic materials into ultrathin format to reduce the stiffness [5, 14], and this is why 2D materials or thin films are widely used [15–16]. However, the thinning-induced flexibility does not change the intrinsic brittleness and rigidity [17–18] of the inorganic semiconductors, which are constituted mainly by covalent or ion-covalent bonding [19].

Regarding this issue, another important mechanical property, plasticity, should be considered. As a matter of fact, however, plasticity is a long-sought target for inorganic materials, e.g., ceramics [20]. On the one hand, plasticity means machinability, that is, plastic ceramics can be mechanically deformed and processed just like metals do. On the other hand, plastic deformation can prevent the sudden, catastrophic, brittle fracture, which is essential to not only structural materials but also functional materials. Hence, the discovery of the room-temperature plastic inorganic semiconductor Ag_2S [21] and the fabrication of full-inorganic Ag_2S-based thermoelectric (TE) power generation modules [22] are ground breaking, opening a new avenue toward next-generation flexible electronics.

This chapter will provide an in-time overview for the newly discovered plastic/flexible inorganic semiconductors. We shall first clarify the concept of flexibility and then illustrate the intrinsic plasticity for metals and brittleness for inorganic materials. Then, we will mention the special plasticity and the chemical bonding origins in a few ionic crystals such as AgCl. After that, we will systematically review the extraordinary mechanical properties of Ag_2S and fully flexible thermoelectric devices. Finally, the prospect and challenge for plastic inorganic semiconductors as flexible electronic materials will be discussed.

2. Fundamental concepts: flexibility

As a matter of fact, "flexibility," unlike "ductility" or "rigidity," is not a scientifically clear concept. For flexible electronics, it is widely conceived that only elastic deformation is needed or allowed. In this sense, "flexibility" refers to the ability of the material/device to bend easily in an elastic way. According to Peng and Snyder [23], flexibility f is quantified by the largest curvature of bending, $1/r_b$, where r_b is the minimum bending radius. A material with a thickness h bent about a radius r_b experiences the greatest tensile and compressive stresses on the outer and inner faces, respectively. The maximum strain is readily calculated from the geometry considering that the middle (neutral layer) of the material is unstrained:

$$\varepsilon = \frac{\left[2\pi\left(r_b + \frac{h}{2}\right) - 2\pi r_b\right]}{2\pi r_b} = \frac{h}{2r_b} \tag{1}$$

The maximum elastic strain is reached when plastic deformation occurs: $\varepsilon = \sigma_y/E$, where E is the elastic (Young's) modulus. Thus, the flexibility becomes:

$$f = \frac{1}{r_b} = \frac{2}{h}\varepsilon_y = \frac{2}{h}\frac{\sigma_y}{E} \tag{2}$$

Removing the shape factor $(2/h)$, the material flexibility is $f_{\mathrm{fom}} = \sigma_y/E$.

The value of f_{fom} is plotted in **Figure 1** for various materials. It is seen that this definition with the plot well distinguishes commonly sensed flexible materials such as rubber between rigid ones like ceramics. In fact, this definition is essentially consistent with the commonly seen bending stiffness or flexural rigidity, $f_r \sim h^3E$ [14, 24]. That is, intrinsically stiff materials with a large thickness tend to be rigid or not flexible. Accordingly, (elastic) flexibility can be realized in thin films or flakes. Therefore, two-dimensional (2D) materials are popular candidates for flexible electronics.

The above definition treats flexibility as elastic deformability, which is reasonable considering the real application. However, as discussed in Section 1, plasticity is important, particularly for efficient material processing. A plastic material can be easily processed into target geometry with little fracture or waste. In addition,

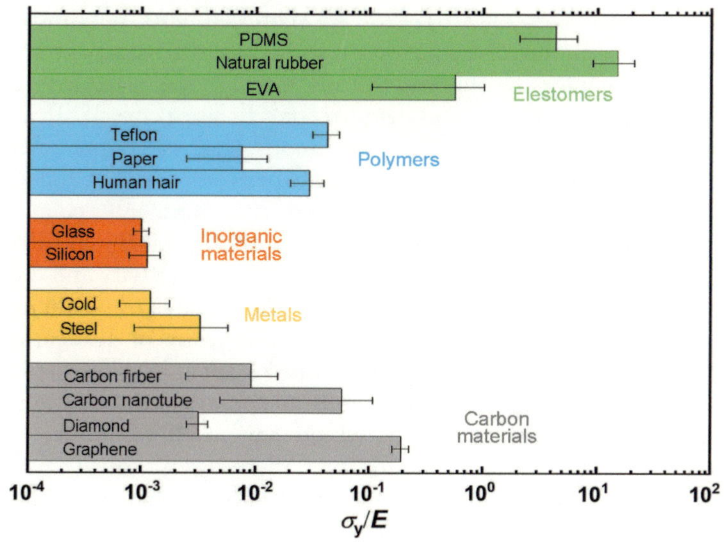

Figure 1.
The ratio of yield strength to Young's modulus for various materials. Raw data are taken from Ref. [23].

a plastic material gains intrinsic flexibility without the restrict of size, that is, it can be readily deformed without breaking even in the bulk format, which is essentially important for applications requiring energy densities. Nonetheless, plasticity is rarely seen in inorganic semiconductors, which will be discussed in Section 3.

3. Prevalent brittleness for inorganic semiconductors

Ductile and brittle behaviors are schematically shown in **Figure 2(a)**. Inorganic semiconductors are mostly constituted by covalent or ion-covalent bonds [25], which assure an appreciable electron orbital overlap, dispersive electronic band, and decent carrier mobility [26]. Covalent bonds are directional, saturated, and localized. Therefore, even a trivial bonding distortion will cause a large instability, which is vividly demonstrated by the deep and steep curve in the interatomic potential versus

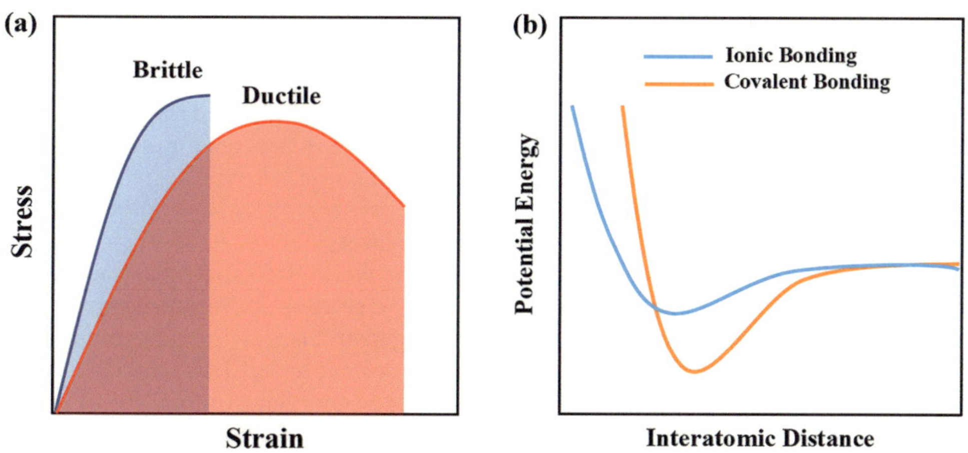

Figure 2.
(a) Stress-strain curves for brittle and ductile materials and (b) atomic potential varying with atomic distances.

atomic distance diagram (**Figure 2(b)**) [20]. As a contrast, ionic bonding is unidirectional and nonsaturated. The columbic force is somewhat diffuse and extended within a large space. Consequently, the bonding distortion will induce a smaller variation in energy, as reflected by the shallow and flat potential-distance curve [20]. Nonetheless, charges in ionic materials are localized around anions, leading to a poor electrical conductivity. Metallic bonding strength is mediate between covalent and ionic bonds. Metal atoms use their outer-shell electrons for two- and multicenter polar interactions, which open the possibility to unite charge transfer with a nonnegligible density of states at the Fermi level. This makes up good electrical conductivity and ductility.

Beyond the covalent bonding characteristics, the absence of plasticity and flexibility in inorganic materials can be interpreted in terms of dislocations. Various defects exist in inorganic materials (e.g., ceramics) such as vacancies, interstitials, and voids, strongly inhibiting the movement of dislocations. In addition, grain boundaries can cause dislocation pile-up, which may be a kind of crack source. For metals, the pile-up of dislocations near grain boundaries will cause hardening in mechanical properties. For ceramics, ductility is limited by crack nucleation and glide. In fact, the main deformation mechanism for most ceramics is creep. This process is mainly controlled by dislocation climb and diffusion, both of which are strongly dependent on temperature.

4. Plastic ionic inorganic materials

Although most inorganic materials are generally brittle, some ionic crystals have been found to exhibit some degree of plasticity upon certain deformation even at room temperature, such as AgCl [27–28], KCl [29], and LiF [30]. Under a slow strain rate, these crystals exhibit tensile properties like metals and show the phenomenon of neck down [29]. This ductile behavior is attributed to the wavy slip [31–32]. During the deformation, each grain has its own change and should conform to the distortion of neighbors [33]. The wavy slipping allows to change planes in the vicinity of grain boundaries to permit distortions. The ability of crystals to relax the stress is essential for such a ductile deformation. At higher temperature, dislocations near the grain boundaries are easier to change the slipping planes due to the thermal active diffusion mentioned above, spreading out from boundaries to another grain. Thus, plastic deformation above transition temperature is mainly caused by the increase in slip systems and the dislocate-diffusion-induced creep and cross-slip.

First-principle calculations have been performed to compare the plasticity between AgCl and NaCl [34]. The generalized stacking fault energy (GSFE) and the double of the surface energy $2\gamma_s$ were used as slipping barrier energy and cleavage energy, respectively. If GSFE > $2\gamma_s$, rupture occurs. It was found that in NaCl, GSFE >$2\gamma_s$ happens before reaching the middle of slipping process for most slipping systems. While for AgCl, interestingly, GSFE remains smaller than $2\gamma_s$ over the whole slipping process for <110>{110}/{111}/{001} systems. It means that the three slipping systems are stable for plastic deformation in AgCl.

The reason for the different plastic behaviors between NaCl and AgCl was further attributed to their fundamental differences in electronic structures (**Figure 3**). The calculations [34] show that the smaller band gap and larger atomic-orbital overlap (Ag-5s and Cl-3p) contribute to a weak ionic bonding in AgCl. Crystal Orbital Hamilton population (COHP) analysis shows that the energies of initial Ag-Cl bonds gradually decrease as slipping proceeds. Particularly, in the (110), (111) planes, new Ag-Cl bonds are formed; a strong Ag-Ag bond is also formed when slipping on (100) planes. These newly formed bonds lower the unstable GSF energies in these slipping systems. While in NaCl, no newly formed bonds are observed.

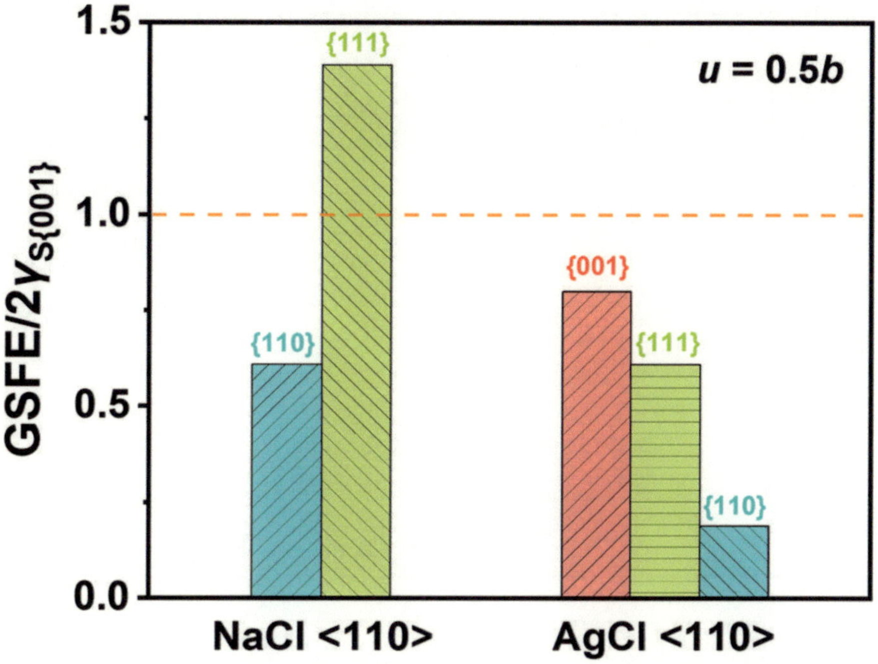

Figure 3.
The ratio of GSF energy to the double surface energy 2γs of the {001} plane for different slipping systems in NaCl and AgCl. u denotes displacement, and b is the Burgers vector. The data are taken from Ref. [34].

The plasticity in AgCl was also understood from the perspective of dislocations [35]. Electronic structures of dislocations in AgCl exhibit larger bonding interactions between atomic orbitals than that of NaCl. Also, the nearest neighbor distances around the dislocation core are tend to be shorter than that in NaCl. The two properties lead to a lower core energy (E_{core}) for AgCl. Far from the core, atoms are displaced elastically, and the energy stored elastically ($E_{elastic}$) for AgCl is also estimated to be lower than that in NaCl. The dislocation excess energies ($E_{elastic} + E_{core}$) in NaCl are at least two times higher than that in AgCl, regardless of the type of dislocation. Therefore, it is expected that dislocations in AgCl are much easier to nucleate and multiply from various sources.

5. Plasticity of ZnS in darkness

Very recently, ZnS, a well-known brittle material, was also reported to exhibit extraordinary "plasticity" in complete darkness [36]. ZnS crystals fractured immediately when they deformed under light irradiation. However, the crystals could be plastically deformed to a compression strain of 45% in complete darkness as shown in **Figure 4**. It is also found that the optical band gap decreased by 0.6 eV after deformation as also clearly reflected by the apparent colors, which is probably due to the formation of extra energy levels at the bandgap edge in the presence of dislocations.

Based on optical and electronic microscopies, the plastic deformation in complete darkness is caused by glide and multiplication of dislocations belonging to the primary slip system. By contrast, plastic deformation under light irradiation involves deformation twinning. Obviously, the latter corresponds to a much poorer plasticity.

The origins for different deformation types are explained below. The dislocations in ZnS decompose into two partial dislocations. In darkness, the synergetic

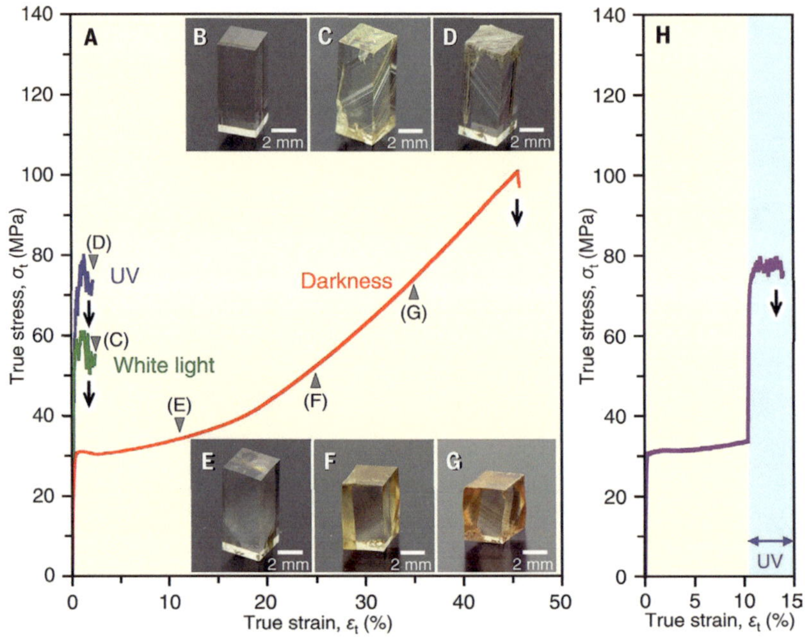

Figure 4.
Characterizations of plastic deformation. (a) Stress-strain curves of ZnS single crystals under white or UV light (365 nm) or in complete darkness. (b) An undeformed specimen. (c and d) The specimens deformed under (c) white light-emitting diode (LED) light and (d) UV LED light (365 nm). (e–g) The specimens deformed up to (e) ε = 11%, (f) ε = 25%, and (g) ε = 35% in complete darkness. (h) A stress-strain curve obtained by a deformation in complete darkness up to ε = 10% and the subsequent deformation under UV light. Adapted from Ref. [36] with the permission from AAAS, Copyright 2019.

glide motion of the two partials will cause large slip deformations. On the contrary, under light irradiation, photo-excited electrons or holes can be trapped, thus charging some dislocations. The mobility of the dislocation can be limited by dragging the surrounding charge cloud compensating the dislocation charge. Therefore, the different charge states of the two dislocations in ZnS will lead to the great difference of their mobility, which will lead to the observed deformation twinning.

The work suggests that the mechanical properties are strongly intercorrelated to optical and electronic properties. It also implies that inorganic semiconductors are not necessarily "intrinsically" brittle.

6. Room-temperature plastic semiconductor Ag_2S

The plastic ionic materials like AgCl are nearly insulating and cannot be used as semiconductors. Recently, α-Ag_2S was discovered as the room-temperature ductile inorganic semiconductor as shown in **Figure 5** [21]. The plasticity of Ag_2S is extraordinary: the engineering strains are ~4.5% in tension, 50% in compression, and above 20% in three-point bending, typical characteristics of metals as shown in **Figure 6**. α-Ag_2S is a typical nondegenerate n-type semiconductor with a low electron carrier concentration about (10^{14}–10^{15}) cm^{-3} and a large, negative Seebeck coefficient (around −1000 μV/K) at room temperature. The band gap is around 1 eV, and RT electrical conductivity ranges from 0.09 to 0.16 Sm^{-1}. The carrier mobility μ_H is around 100 cm^2/Vs [21].

The marvelous plasticity of α-Ag_2S comes from its special crystal structure. α-Ag_2S adopts a monoclinic symmetry with the space group $P2_1/c$, consisting of

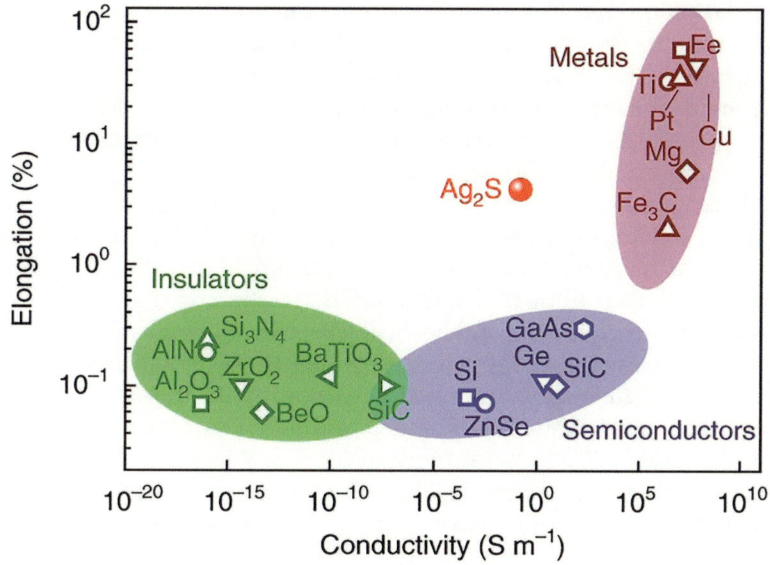

Figure 5.
Elongation versus electrical conductivity for α-Ag_2S and various materials. Adapted from Ref. [21] with permissions from Springer Nature, Copyright 2018.

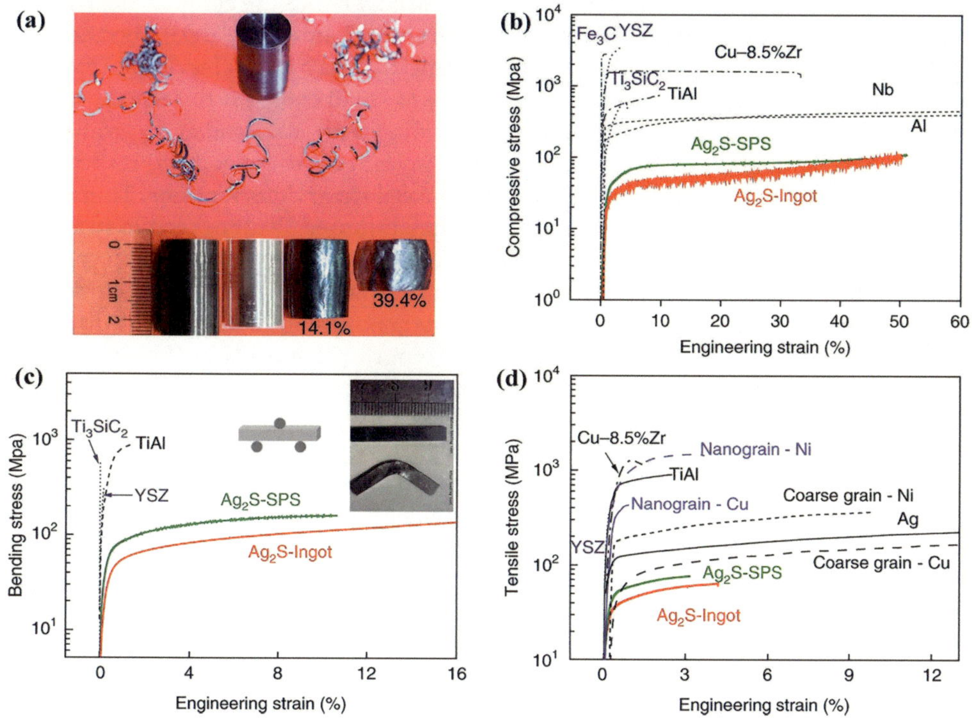

Figure 6.
Room-temperature mechanical properties of the semiconductor α-Ag_2S. (a) A machined cylinder for the compression test (top) and its deformations under hammering (bottom). (b–d) Strain-stress curves for compression (b), bending, (c) and tension (d) tests at room temperature. Typical materials such as the ceramics yttria-stabilized zirconia (YSZ) and Ti, SiC; the metals Al, Nb, Ni, Cu, Ag, Cu–8.5%Zr alloy, and Fe_3C; and the intermetallic compound TiAl are shown for comparison. The inset in c shows the as-cast ingot samples before and after the bending test. Adapted from Ref. [21] with permissions from Springer Nature, Copyright 2018.

tetramolecular units. Four S and four Ag atoms constitute eight-atomic ring fragments interlinked by the sulfur atoms. Precisely among (100) plane, a wrinkled structure formed by two S and six Ag atoms stacking along [100] direction, and this structure provides channels for slipping. In addition, it was found that Ag sites were occupied only for 70%, while S sites occupied completely. These unfixed Ag atoms may induce additional Ag-S and Ag-Ag bonds.

The multicentered, diffuse, and relatively weak bonding gives rise to the small slipping energy and large cleavage energy, i.e., plastic material can slip easily without cleavage as shown in **Figure** 7. As for α-Ag_2S, it is assumed that slip plane is (100) and slip direction [001]. According to the calculation, the slipping energy (E_B) is 150 meV per atom for Ag_2S, which is comparably small with conventionally ductile metals (Ti, Mg); also, the cleavage energy is relatively large for Ag_2S, which is 148 meV per atom, indicating certain relatively strong forces interlinking those slip planes instead of cleaving. In comparison, the values are, respectively, less than 60 meV for NaCl, graphite, and diamond; 570 meV for Mg; and 2150 meV per atom for Ti.

The distribution of the electron localizability indicator (ELI-D) shows a local maximum on the outer side of each S atom, and the basin of this maximum is caused by the formation of a lone pair or a strong Ag-S interaction shown in quantum theory of atoms in molecules (QTAIM). Besides, these lone pairs form double layers in the (100) plane. Thus, the E_B is supposed to be small due to a relative weak interaction between the lone pairs. The COHP calculations reveal chemical bonds changing during the glide. During the slipping process, some bonds vanish, while new bonds form continuously during the whole process, and these new bonds are comparable with the Ag-S bonds between layers in strength. Therefore, S atoms are always bonded with Ag atoms, resulting in a large ΔE_C to prevent materials from cleaving. In short, S atoms move along Ag-formed tracks easily due to small energy difference in steps, while these are difficult to cleavage for tight bonding with surrounded Ag atoms.

Li et al. [37] applied *ab initio*-based DFT to investigate the structural response of Ag_2S under pure shear, uniaxial tension, and biaxial shear deformations. To simulate quasi-static mechanical loading process, they imposed shear or tensile strain on a particular system gradually. It is found that Ag_2S has a theoretical shear strength of 1.02 GPa in the (001) [010] and (100) [010] slip systems (**Figure 8(a)**) lower than that of common metal and ceramics. The ideal tensile strength along [100] direction is 2.2 GPa, significantly higher than its shear strength (**Figure 8(b)**). The very low shear strength under these slip systems is expected to create pathways for easy slip.

The Ag-S octagon structure would be highly distorted under shear deformation. However, the Ag-S bond lengths change only slightly during shearing. Only when the strain is larger than ~110%, bond breakage would happen. This feature implies that Ag_2S would retain its structural integrity even under very large shear deformation. Under (100) [010] shear load, Ag-Ag metallic bonds would form at 0.671 strain. These newly formed bonds strengthen the Ag-S frameworks and contribute to its structural integrity. Under [100] tensile load, the processes of breakage and the formation of Ag-S bond happen at the same time. The structural integrity could thus be preserved to a large strain. Based on these discoveries, they proposed that the easy slip pathways with good structural integrity under shear load are the origin of ductility in Ag_2S.

Recent research also found that the single-layer Ag_2S is a kind of auxetic materials with unusual negative Poisson's ratio [38]. According to their calculations, single-layer Ag_2S has relatively low Young's modulus with 61.61 $N \cdot m^{-1}$ along [010] direction and 2.78 $N \cdot m^{-1}$ along [100] direction. This is the lowest value found in various 2D materials such as graphite and MoS_2. This means that along [100] direction, single-layer Ag_2S will show extraordinary flexibility. Moreover, single-layer Ag_2S shows negative Poisson's ratio in both in-plane and out-of-plane directions, which is unique among quasi-2D materials.

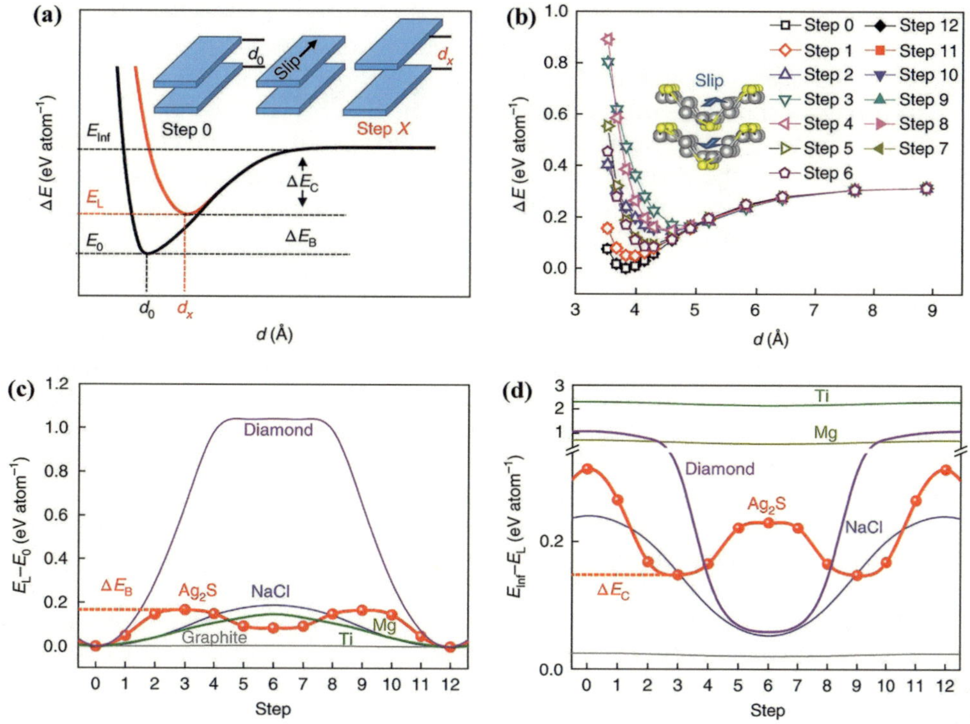

Figure 7.
(a) Crystal structure of α-Ag_2S along [001] direction. (b) Schematic map for energy variation as a function of interlayer distance d *during slipping. (c) ($E_L - E_o$) and (d) ($E_{Inf} - E_L$) behavior during slipping. Adapted from Ref. [21] with permissions from Springer Nature, Copyright 2018.*

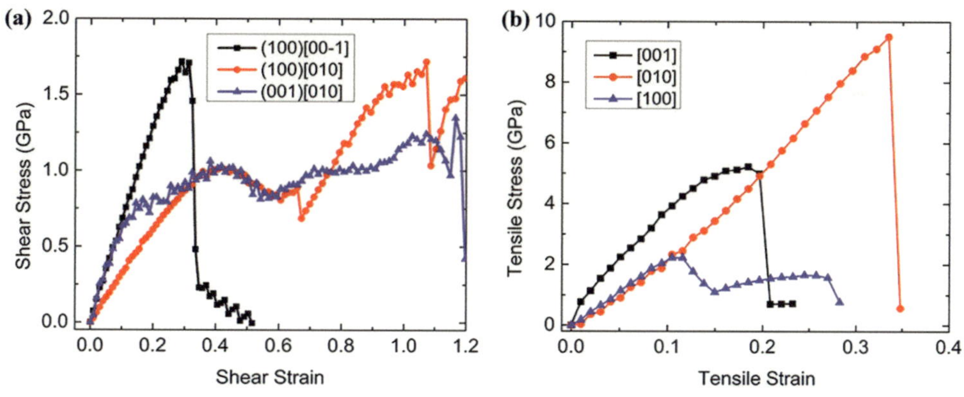

Figure 8.
Stress response of α-Ag_2S against pure shear strain and biaxial tensile strain, respectively. (a) Shear-stress-shear-strain relations along various slip systems. (b) Tensile-stress-tensile-strain relations along various tensile systems. Adapted from Ref. [37] under the Creative Commons CC BY license.

7. Full inorganic flexible thermoelectric materials

The discovery of plastic Ag_2S makes it possible to fabricate full-inorganic flexible devices. Pristine Ag_2S exhibits a medium band gap (~1.0 eV), high mobility (~100 cm^2/Vs), and extremely low lattice thermal conductivity (~0.5 $Wm^{-1}K^{-1}$ near room temperature) [21], which makes it a potential candidate for thermoelectric application. However, the extremely low carrier concentration (about 1.6 × 10^{14} cm^{-3} at room temperature) leads to its poor electrical conductivity. Therefore, the thermoelectric performance of Ag_2S should be further optimized, and its plasticity should be maintained at the same time.

By alloying with Se/Te, Liang et al. successfully tuned the carrier concentration of Ag_2S by virtue of the lowered defect formation energy of Ag interstitial atoms. Consequently, the electrical conductivity and power factor were largely optimized: the electrical conductivities of $Ag_2S_{0.5}Se_{0.5}$, $Ag_2S_{0.8}Te_{0.2}$, and $Ag_2S_{0.5}Se_{0.45}Te_{0.05}$ reach around 10^4 S m^{-1} at room temperature, which are comparable to state-of-the-art brittle inorganic TE materials. The power factors of Ag_2S-based materials can reach 5 μW·cm^{-1}·K^{-2} at room temperature. Meanwhile, Ag_2(S, Se), Ag_2(S, Te), and Ag_2(S, Se, Te) have record low thermal conductivities of 0.3~0.6 Wm^{-1} K^{-1} at 300–450 K, among the lowest values observed in fully densified inorganic solids. The band gap of Se or Te alloyed Ag_2S-based materials is also reduced, shifting the peak value of zT toward a lower temperature. Thus, a highest zT value about 0.44 could be realized for $Ag_2S_{0.5}Se_{0.45}Te_{0.05}$ at room temperature.

More interestingly, alloying does not severely impair the plasticity. According to the mechanical property test, the ductility and flexibility of Ag_2S-based materials would be maintained if the Se content is less than 60% or the Te content is less than 70%. Hence, the materials would possess both good ductility and TE performance when the Se/Te content is in the range of 20~60%, as shown in **Figure 9(c)**. To further verify the robustness of Ag_2S-based materials under different usage scenarios, bending cycle test was conducted on $Ag_2S_{0.5}Se_{0.5}$ strip with a thickness of about 10 μm. As shown in **Figure 9(f)**, the variation in Seebeck coefficient and electrical conductivity is less than 10% after 1000 bending cycles with a bending radius less than 3 mm. In addition, the relative resistance variation in the strip under different bending radius is also estimated.

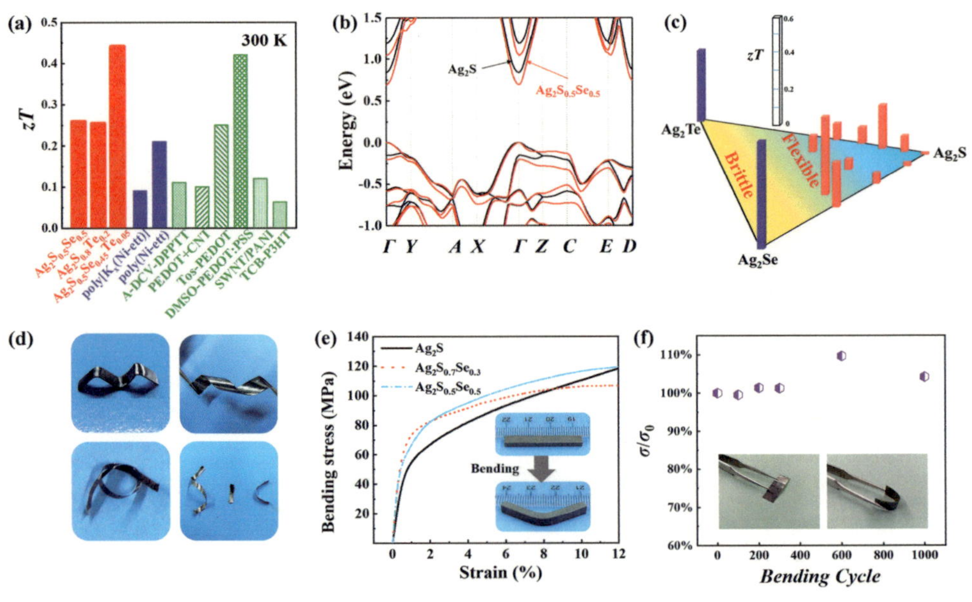

Figure 9.
(a) TE figure of merit zT for Ag_2(S, Se), Ag_2(S, Te), and Ag_2(S, Se, Te) at 300 K. Several representative organic TE materials are included for comparison. (b) Calculated band structure for Ag_2S and $Ag_2S_{0.5}Se_{0.5}$. (c) Flexibility-zT phase diagram of Ag_2S-Ag_2Se-Ag_2Te system. (d) The $Ag_2S_{0.5}Se_{0.5}$ and $Ag_2S_{0.8}Te_{0.2}$ samples twisted to various shapes. (e) Bending tests of $Ag_2S_{1-x}Se_x$ (x = 0, 0.1, 0.3, 0.5). (f) Relative electrical conductivity variation σ/σ_0 of the $Ag_2S_{0.5}Se_{0.5}$ foil after various number of times of bending cycles. Reproduced from Ref. [22] with permissions from The Royal Society of Chemistry. Copyright 2019.

Using $Ag_2S_{0.5}Se_{0.5}$ strips as n-type legs and Pt-Rh wires as p-type legs, full-inorganic thermoelectric devices were fabricated as shown in **Figure 10** [22]. Under temperature difference 20 K, the maximum power of the device is 10 μW, and the normalized maximum power density $P_{max}L/A$ reaches 0.08 $W \cdot m^{-1}$, which is much higher than those for current organic TE devices. This research has solved the most fundamental and challenging problems for the fabrication of full-inorganic and high-performance flexible thermoelectrics, opening a new direction for inorganic flexible thermoelectrics.

The conventional strategy toward flexible TE devices is mounting TE thin-film materials onto the intrinsically flexible substrates. Ding et al. [39] fabricated flexible TE devices based on Ag_2Se nanowires and plastic nylon substrate. The hybrid film was hot pressed at 200°C and 1 MPa for 30 min, which endows the film high TE performance and excellent flexibility at the same time. The highest power factor value of as-prepared Ag_2Se/nylon film reaches 9.87 $\mu W m^{-1} K^{-2}$ at 300 K, almost the highest value among reported n-type flexible TE materials. Although the internal resistance of the four legs of Ag_2Se/nylon devices is up to 250 Ω, the in-plane TE device exhibits a high-power density of 2.3 W m^{-2} under a temperature difference of 30 K.

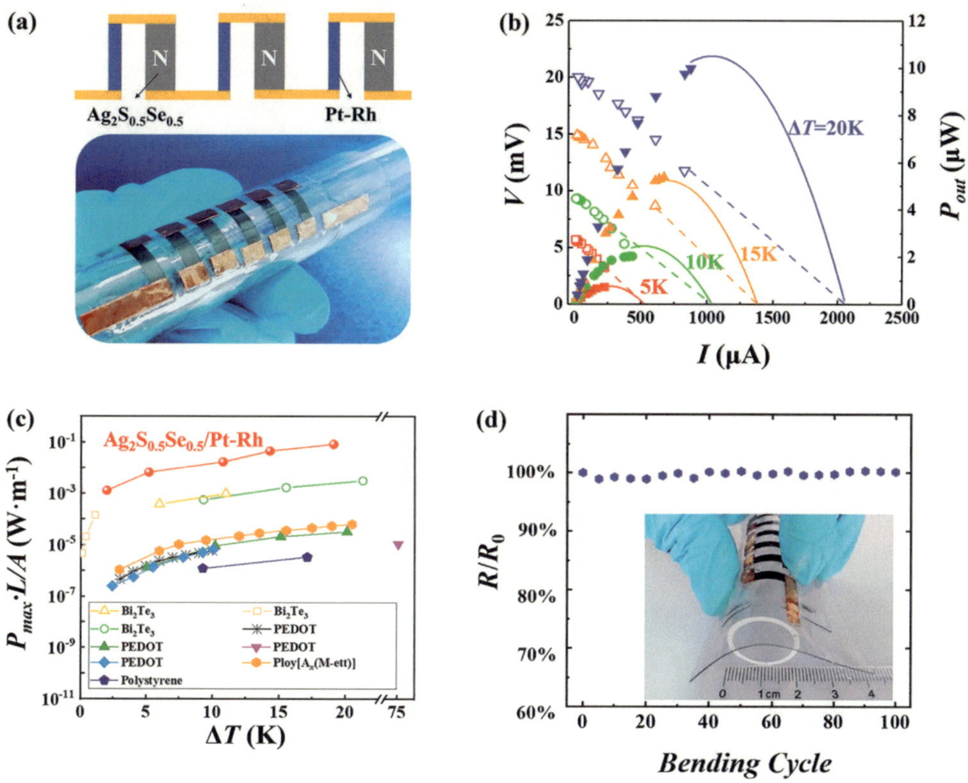

Figure 10.
(a) Upper panel: A schematic of the $Ag_2S_{0.5}Se_{0.5}$/Pt-Rh in-plane device with $Ag_2S_{0.5}Se_{0.5}$ as n*-type legs and Pt-Ru wire as* p*-type legs. Bottom panel: Optical image of a six-couple flexible $Ag_2S_{0.5}Se_{0.5}$/Pt-Rh TE device. The as-shown in-plane device is merely for the purpose of demonstration. (b) Output voltage* V *and power output* P_{out} *as a function of current* (I) *for a six-couple $Ag_2S_{0.5}Se_{0.5}$/Pt-Rh device under different operating temperature differences. The cold side temperature is fixed at 293 K. (c) Comparison of normalized maximum power density* ($P_{max}L/A$) *among the $Ag_2(S, Se)$-based inorganic TE device, inorganic-organic hybrid flexible TE devices, and organic flexible TE devices. (d) Relative electrical resistance variation* R/R_0 *of the $Ag_2S_{0.5}Se_{0.5}$/Pt-Rh TE device after bending various number of times. The inset shows the optical image for the bended device. The bending radius is 10 mm. Reproduced from Ref. [22] with permissions from The Royal Society of Chemistry. Copyright 2019.*

8. Summaries and outlook

This chapter reviews the newly emerging plastic inorganic semiconductors (e.g., Ag_2S) for next-generation flexible electronics. The term "flexibility" is clarified at the very beginning. It should be recognized that plasticity is important for flexible electronics due to the availability of feasible processing and deformability free of size restricts. The intrinsic brittleness for inorganic semiconductors and ceramics is explained from unidirectional and saturated characteristics of covalent bonds. Historically, ionic crystals like AgCl have been found to exhibit certain plasticity but lack decent electrical conductivity. Groundbreakingly, Ag_2S was discovered as the room-temperature ductile semiconductor. From the chemical bonding perspective, the multicentered, diffuse, weak interactions induce easy slipping while maintaining the integrity, which holds well not only for Ag_2S but also for other plastic materials. The generalized bonding features are useful guidance for developing new flexible/plastic semiconductors. The electrical properties and thermoelectric performance of Ag_2S are readily optimized upon Se/Te alloying while maintaining the plasticity and flexibility. Successively, full-inorganic thermoelectric devices are fabricated based on plastic and flexible Ag_2S-based semiconductors, yielding much higher output power density than organic counterparts.

The discovery and application demonstration of plastic Ag_2S inorganic semiconductor pave a new way toward next-generation flexible electronics. Facing large-scale applications in electronics and energy conversions, several key challenges lie ahead. First and basically, the mechanisms for plastic deformation in Ag_2S needs further investigation, especially on the individual and synergetic effects of both chemical bonding and dislocations, which calls for tremendous efforts of both experimentalists and theorists from a variety of disciplines. Second, practical criteria are required to rapidly screen potentially new, plastic/flexible semiconductors. These performance indicators should be easily available yet insightful, and it is better that they can be implemented into the high-throughput calculations. Third, all the techniques are to be renewed including material processing, electrode/substrate selection, device fabrication, and circuit integration.

Facing all these exciting challenges and fascinating opportunities, there is no doubt that the flexible/plastic inorganic semiconductors will bring a revolution to academic communities, electronic/energy industries, and worldwide market. The next-generation flexible electronics is meant to deeply change our life and reshape the world. The future has come.

Author details

Tian-Ran Wei[1,2], Heyang Chen[2], Xun Shi[1,2] and Lidong Chen[1*]

1 State Key Laboratory of High Performance Ceramics and Superfine Microstructure, Shanghai Institute of Ceramics, Chinese Academy of Sciences, Shanghai, China

2 State Key Laboratory of Metal Matrix Composites, School of Materials Science and Engineering, Shanghai Jiao Tong University, Shanghai, China

*Address all correspondence to: cld@mail.sic.ac.cn

References

[1] Wong WS, Salleo A. Flexible Electronics: Materials and Applications. Vol. 11. New York: Springer Science & Business Media; 2009

[2] Kim D-H, Lu N, Ma R, Kim Y-S, Kim R-H, Wang S, et al. Epidermal electronics. Science. 2011;**333**(6044): 838-843

[3] Gates BD. Flexible electronics. Science. 2009;**323**(5921):1566-1567

[4] Sun Y, Rogers JA. Inorganic semiconductors for flexible electronics. Advanced Materials. 2007;**19**(15):1897-1916

[5] Cavallo F, Lagally MG. Semiconductors turn soft: Inorganic nanomembranes. Soft Matter. 2010;**6**(3):439-455

[6] Rogers JA, Lagally MG, Nuzzo RG. Synthesis, assembly and applications of semiconductor nanomembranes. Nature. 2011;**477**(7362):45-53

[7] Kaltenbrunner M, Sekitani T, Reeder J, Yokota T, Kuribara K, Tokuhara T, et al. An ultra-lightweight design for imperceptible plastic electronics. Nature. 2013;**499**(7459): 458-463

[8] Zhang S, Xu X, Lin T, He P. Recent advances in nano-materials for packaging of electronic devices. Journal of Materials Science: Materials in Electronics. 2019;**30**:13855-13868

[9] Yang M, Kim SW, Zhang S, Park DY, Lee C-W, Ko Y-H, et al. Facile and highly efficient fabrication of robust Ag nanowire–elastomer composite electrodes with tailored electrical properties. Journal of Materials Chemistry C. 2018;**6**(27):7207-7218

[10] Xu J, Wang S, Wang G-JN, Zhu C, Luo S, Jin L, et al. Highly stretchable polymer semiconductor films through the nanoconfinement effect. Science. 2017;**355**(6320):59-64

[11] Sekitani T, Yokota T, Zschieschang U, Klauk H, Bauer S, Takeuchi K, et al. Organic nonvolatile memory transistors for flexible sensor arrays. Science. 2009;**326**(5959):1516-1519

[12] Oh JY, Rondeau-Gagne S, Chiu YC, Chortos A, Lissel F, Wang GN, et al. Intrinsically stretchable and healable semiconducting polymer for organic transistors. Nature. 2016;**539**(7629):411-415

[13] Jin ML, Park S, Kim JS, Kwon SH, Zhang S, Yoo MS, et al. An ultrastable ionic chemiresistor skin with an intrinsically stretchable polymer electrolyte. Advanced Materials. 2018;**30**(20):1706851

[14] Timoshenko SP, Woinowsky-Krieger S. Theory of Plates and Shells. Auckland: McGraw-hill; 1959

[15] Bhimanapati GR, Lin Z, Meunier V, Jung Y, Cha J, Das S, et al. Recent advances in two-dimensional materials beyond graphene. ACS Nano. 2015;**9**(12):11509-11539

[16] Fiori G, Bonaccorso F, Iannaccone G, Palacios T, Neumaier D, Seabaugh A, et al. Electronics based on two-dimensional materials. Nature Nanotechnology. 2014;**9**:768-779

[17] Lee C, Wei X, Kysar JW, Hone J. Measurement of the elastic properties and intrinsic strength of monolayer graphene. Science. 2008;**321**(5887):385-388

[18] Kim J, Kim J, Song S, Zhang S, Cha J, Kim K, et al. Strength dependence of epoxy composites on the average filler size of non-oxidized graphene flake. Carbon. 2017;**113**:379-386

[19] Faber KT, Malloy KJ. The Mechanical Properties of Semiconductors. San Diego: Academic Press; 1992

[20] Green DJ. An Introduction to the Mechanical Properties of Ceramics. Cambridge: Cambridge University Press; 1998

[21] Shi X, Chen H, Hao F, Liu R, Wang T, Qiu P, et al. Room-temperature ductile inorganic semiconductor. Nature Materials. 2018;**17**:421-426

[22] Liang J, Wang T, Qiu P, Yang S, Ming C, Chen H, et al. Flexible thermoelectrics: From silver chalcogenides to full-inorganic devices. Energy & Environmental Science. 2019;**12**(10):2983-2990

[23] Peng J, Snyder GJ. A figure of merit for flexibility. Science. 2019;**366**(6466):690-691

[24] Lu Q, Arroyo M, Huang R. Elastic bending modulus of monolayer graphene. Journal of Physics D: Applied Physics. 2009;**42**(10):102002

[25] Yu PY, Cardona M. Fundamentals of Semiconductors. Berlin: Springer; 2010

[26] Zeier WG, Zevalkink A, Gibbs ZM, Hautier G, Kanatzidis MG, Snyder GJ. Thinking like a chemist: Intuition in thermoelectric materials. Angewandte Chemie, International Edition. 2016;**55**(24):2-18

[27] Nye JF. Plastic deformation of silver chloride I. internal stresses and the glide mechanism. Proceedings of the Royal Society of London, Series A: Mathematical, Physical and Engineering Sciences. 1949;**198**(1053):190-204

[28] Nye JF. Plastic deformation of silver chloride. II. Photoelastic study of the internal stresses in glide packets. Proceedings of the Royal Society of London. Series A: Mathematical and Physical Sciences. 1949;**200**(1060):47-66

[29] Stoloff N, Lezius D, Johnston T. Effect of temperature on the deformation of KCl-KBr alloys. Journal of Applied Physics. 1963;**34**(11):3315-3322

[30] Taylor A, Albers H, Pohl R. Effect of plastic deformation on the thermal conductivity of various ionic crystals. Journal of Applied Physics. 1965;**36**(7):2270-2278

[31] Skrotzki W, Frommeyer O, Haasen P. Plasticity of polycrystalline ionic solids. Physica Status Solidi. 1981;**66**(1):219-228

[32] Vávra F, Ševčík Z. Formation of wavy slip bands in AgCl crystals at low temperatures. Czechoslovak Journal of Physics B. 1986;**36**(4):509-513

[33] Stokes R, Li C. Dislocations and the strength of polycrystalline ceramics. In: Stadelmaier HH, Austin WW, editors. Materials Science Research. Boston: Springer; 1963. pp. 133-157

[34] Nakamura A, Ukita M, Shimoda N, Furushima Y, Toyoura K, Matsunaga K. First-principles calculations on slip system activation in the rock salt structure: Electronic origin of ductility in silver chloride. Philosophical Magazine. 2017;**97**(16):1281-1310

[35] Ukita M, Nakamura A, Yokoi T, Matsunaga K. Electronic and atomic structures of edge and screw dislocations in rock salt structured ionic crystals. Philosophical Magazine. 2018;**98**(24):2189-2204

[36] Oshima Y, Nakamura A, Matsunaga K. Extraordinary plasticity of an inorganic semiconductor in darkness. Science. 2018;**360**(6390):772-774

[37] Li G, An Q, Morozov SI, Duan B, Goddard WA, Zhang Q, et al.

Ductile deformation mechanism in semiconductor α-Ag_2S. npj Computational Materials. 2018;**4**(1):44

[38] Peng R, Ma Y, He Z, Huang B, Kou L, Dai Y. Single-layer Ag2S: A two-dimensional bidirectional Auxetic semiconductor. Nano Letters. 2019;**19**(2):1227-1233

[39] Ding Y, Qiu Y, Cai K, Yao Q, Chen S, Chen L, et al. High performance n-type Ag_2Se film on nylon membrane for flexible thermoelectric power generator. Nature Communications. 2019;**10**(1):841